IMAGES
of America

FIRE LOOKOUTS OF OREGON

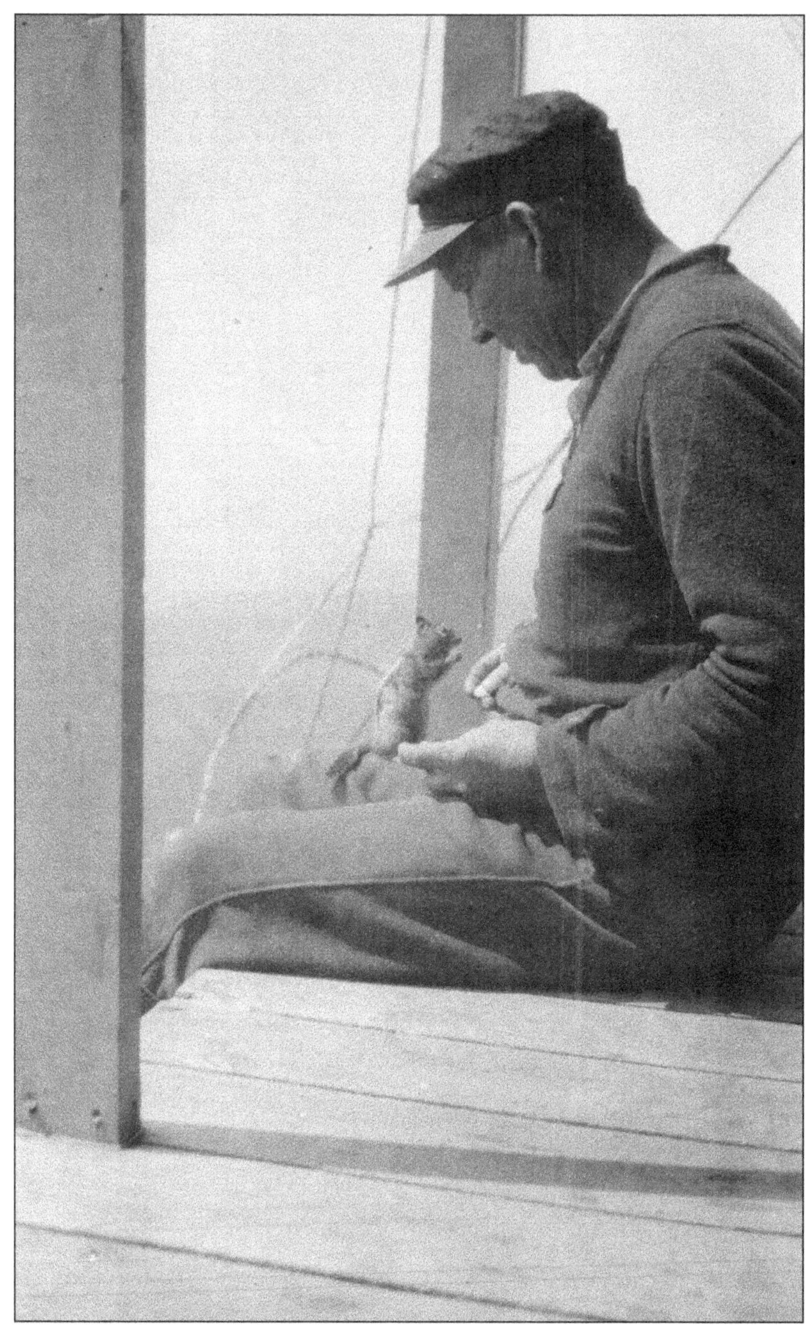

An unidentified man feeds a ground squirrel from the catwalk of a lookout. Some lookouts brought pets with them, but others often made friends with the local wildlife, especially ground squirrels, which are notorious beggars.

On the Cover: The Bull of the Woods lookout was built in 1942. The 10-foot-tall tower now sits at the heart of the Bull of the Woods Wilderness, which was created in 1984. It is the oldest of the seven lookouts still standing in the Mount Hood National Forest, although it is no longer staffed.

IMAGES
of America

Fire Lookouts of Oregon

Cheryl Hill

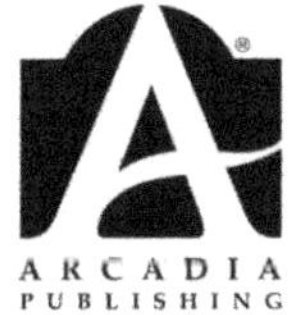

ISBN 978-1-4671-3486-6

Published by Arcadia Publishing
Charleston, South Carolina

Library of Congress Control Number: 2015947044

For all general information, please contact Arcadia Publishing:
Telephone 843-853-2070
Fax 843-853-0044
E-mail sales@arcadiapublishing.com
For customer service and orders:
Toll-Free 1-888-313-2665

Visit us on the Internet at www.arcadiapublishing.com

This book is dedicated to my parents, who nurtured my love of history.

Contents

Acknowledgments

There are many lookout experts and enthusiasts out there, and I have been lucky enough to enlist the help of some of them. Thank you to Ray Kresek, Ron Kemnow, and Don Allen. Thank you also to La Vaughn Kemnow for your editing expertise.

Thank you to the history keepers at the national forest offices and other locations who took time out of their schedules to unearth boxes and albums of old photographs for me to scan and who shared their knowledge with me. Thank you to Debra Barner, Jacob Barnett, Denise Berkshire, Penni Borghi, Kevin Bruce, Sarah Crump, Chris Friend, Cathy Lindberg, Matthew Mawhirter, Patty McNamee, Jen Warren, and Alexandra Wenzl.

One of the joys of this project has been talking to the numerous people who were willing to share their stories with me. Thank you to all the people who spoke to me about your lookout experiences or those of your loved ones. I was not able to use all of your stories in the book, but I thank you all the same. Thank you to C. Rod Bacon, Roger Brandt, Lois Christiansen Eagleton, Annette Finger, Mary Frost, Bob Hanshaw, Donna Henderson, Randall Henderson, Mary Beth Hurlocker, Paul Ingram, Lee Jacobs, Gary Johanson, Donavin Leckenby, Patti Luse, Dee Lynch, Fred Merten, Richard Moles, John-Erik Nilsson, Bill Noland, Lindy Oleson, Benny Parmele, Dan Pinson, Jancy Potterf, Linda Shaw, Al Sorseth, Vicki Timm, and George Wynn.

Unless otherwise noted, all images in this book appear courtesy of the US Forest Service. Other images appear courtesy of the Oregon Department of Forestry (ODF), the Lane County Historical Museum (LCHM), the National Archives, the Forest History Society, Ron Kemnow, George M. Henderson, Bill Noland, Patti Luse, Linda Shaw, and the author's personal collection.

Wherever possible, I have named the people who appear in these images; however, most of the photographs did not include identification.

Introduction

When immigrants started settling the West in the last half of the 19th century, they began inhabiting vast forests. Millions of acres of trees were free for the taking. The 1891 Forest Reserve Act allowed the creation of forest reserves, much to the dismay of timber barons, mine owners, and railroad companies who felt they should be able to do what they wanted with the land. In 1905, the Bureau of Forestry became the US Forest Service, a federal agency that was now in charge of 60 million acres of forest.

The summer of 1910 was hot and dry. In August, numerous small wildfires in Idaho joined together to create what later became known as the Big Blowup. The young Forest Service lacked the funding and manpower to properly fight the fire, and a lack of trails and roads made access to the fire line difficult. When the fire was over, the eastern third of the town of Wallace had burned down, and three million acres of forest were torched. Records vary, but about 85 people were killed. Oregon did not escape unscathed that year. Fires burned all over the state, including a fire that burned more than 60,000 acres in the Crater National Forest (now called the Rogue River–Siskiyou National Forest).

In the wake of that bad year, the Forest Service adopted a policy of strict fire suppression. In each national forest, a fire detection system was organized consisting of lookouts and guard stations. Early communication between these stations was rudimentary, involving heliographs, flags, or even carrier pigeons. These communication methods were soon replaced by a telephone system, which entailed hundreds of miles of No. 9 telephone wire strung throughout the forest. Maintaining these lines was a major job.

In 1911, Forest Service employee William "Bush" Osborne developed an alidade device, which would become known as the Osborne Fire Finder. It sat upon a stand in the center of the lookout and allowed observers to easily take a directional bearing (azimuth) if they spotted smoke. Using known landmarks near the smoke, they could report its approximate location. The Osborne Fire Finder was simple and effective and is still used today in fire lookouts.

Early lookouts were sometimes primitive, often just a crow's nest high up a tree or a fire finder affixed to a mountaintop stump. Later, several different styles of lookouts were developed. They ranged from ground cabins to 100-foot-tall towers. There were D-6 cupola cabins, R-6 cabins, and steel towers with cramped seven-by-seven-foot cabs on top. One of the most common lookout styles was the L-4 model, a 14-foot square frame cabin that was bundled in kits and hauled to the site via pack trains. It was adaptable enough that it could sit on a foundation on the ground or perch atop a tower.

The men and women who were stationed at fire lookouts were also known as lookouts. Their primary duty was to regularly conduct a "check look," inspecting their entire territory for smoke. Lookouts also checked the weather and reported back to headquarters. When they were not on duty or when the weather was bad, they caught up on chores, such as building maintenance, trail maintenance, chopping firewood, washing windows, and hauling water from the nearest spring, lake, or creek.

Starting around 1918, women started staffing lookouts alone. Many were schoolteachers or college students on summer break. Such a job being held by a woman was shocking to some at that time. A 1920 *Oregonian* article called the job "harrowing" for "delicate women." But with billions of dollars worth of timber to protect from wildfire, the Forest Service was willing to hire able-bodied women to staff lookouts.

The 1930s was a boom decade for fire lookout construction. Some lookouts were built on private timberland and by the National Park Service, but most were built by the Forest Service. With the help of the Civilian Conservation Corps, the Forest Service built miles of new trails and roads to peaks and then packed or drove supplies to the summits for the lookouts. A vast network of lookouts sprang up during the Depression.

During World War II, fire lookouts briefly served a purpose other than identifying fires. The US Army established the Aircraft Warning Service (AWS) in 1941 with the goal of spotting enemy aircraft in American airspace and reporting it before significant damage could be inflicted. The system comprised more than 500 observation sites in Oregon, staffed by people who were not fit for service in the regular Army. This resulted in many women serving with the AWS. Observation posts had to be on alert all hours of the day and night year-round, so they were staffed in shifts. Many lookout sites doubled as AWS sites until the program was discontinued in 1944.

Lookout jobs appealed to people from all walks of life—from writers to hermits to teachers—but it was not for everyone. The solitude was attractive to some but felt isolating to others. In addition, living on a mountaintop was a rustic experience, with no plumbing, no electricity, and a hike required to reach the nearest water source.

By the 1950s, patrol planes and helicopters became the preferred method of spotting fires, and the use of lookouts began to decline. In later decades, remote cameras also replaced lookouts. The postwar road-building boom in the national forests resulted in improved access to once-remote areas, enabling crews to respond faster than before. In the 1960s, the Forest Service became worried about liability issues if visitors were to injure themselves on one of the many abandoned and dilapidated structures, so they started removing hundreds of lookouts. Ironically, the preferred method of removal was to burn them down.

Some lookouts survived, however, and their use for fire-spotting never completely stopped. There are less than 175 lookouts left in Oregon, about half of which are still staffed. Today's lookouts are luxurious compared to their historic counterparts. Many are outfitted with solar panels. Gas-powered stoves and refrigerators make mealtime easier, and gas-powered or electric heaters remove the need for chopping and hauling firewood. Communication is achieved by radios, and many lookouts have full cell phone signals.

Almost 20 lookouts in Oregon have found new life as rentals available to the public, a program that started in the 1990s and that has proven to be wildly popular. The National Historic Lookout Register was established in 1990 to recognize historic fire lookouts, and nonprofit organizations such as the Forest Fire Lookout Association and the Sand Mountain Society have put in thousands of volunteer hours to restore and preserve them. Thanks to these efforts, the importance of saving these historic structures has received growing public awareness.

One

North Coast

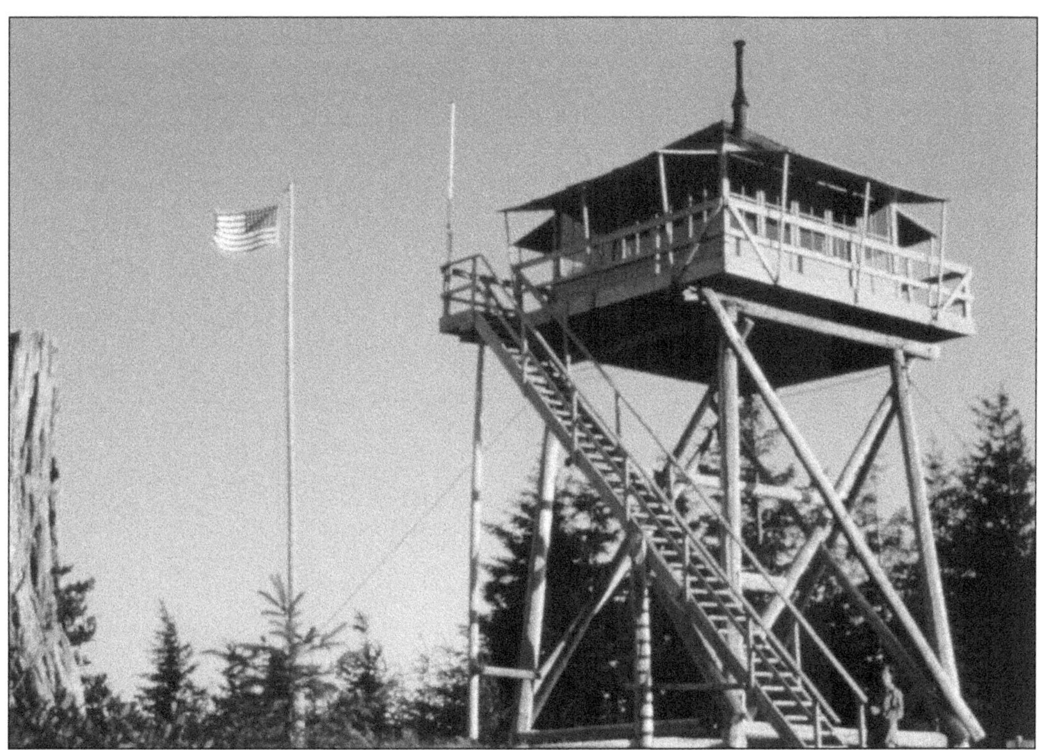

This 15-foot-tall tower was built on Herman Peak in Lane County in the 1930s. When Donald and Dessa Johnson worked here in 1936, they brought Scrappy the kitten with them. They also kept two pet chipmunks who liked to collect cotton, paper scraps, and hair from around the lookout with which to build a nest in their pen (a repurposed box). In 1976, the Forest Service dismantled the tower. The cab was rebuilt atop Iron Mountain in the Willamette National Forest.

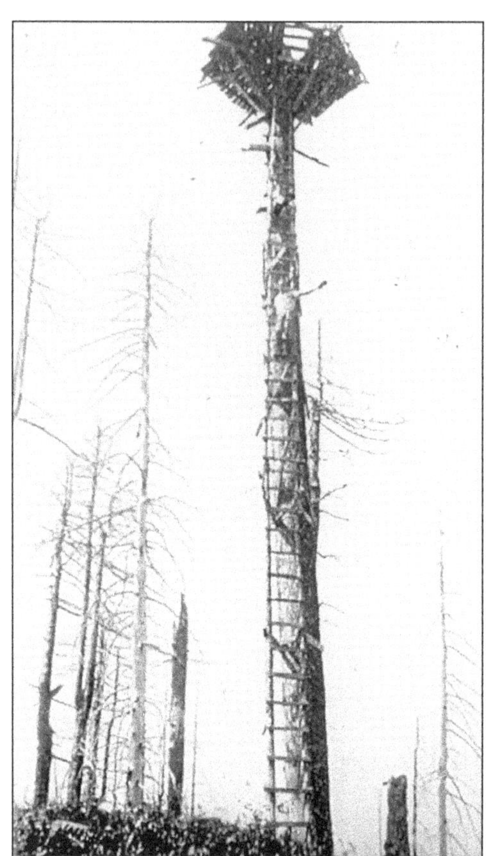

The original crow's nest lookout (pictured) on Cedar Butte was replaced by a 40-foot-tall tower in 1937. While working in that tower in August 1939, Gene Scott found himself surrounded by flames from the Trask Fire when the winds shifted. He managed to escape by swimming and wading four miles down the Wilson River. (ODF.)

Oregon State University student Bill Ruhmann spent his summer break working on Roman Nose Mountain in 1926. On a hot afternoon, he reported smoke east of Cottage Grove but later heard back from the dispatcher that the smoke was from two wood-burning locomotives being used in the filming of the Civil War movie *The General.*

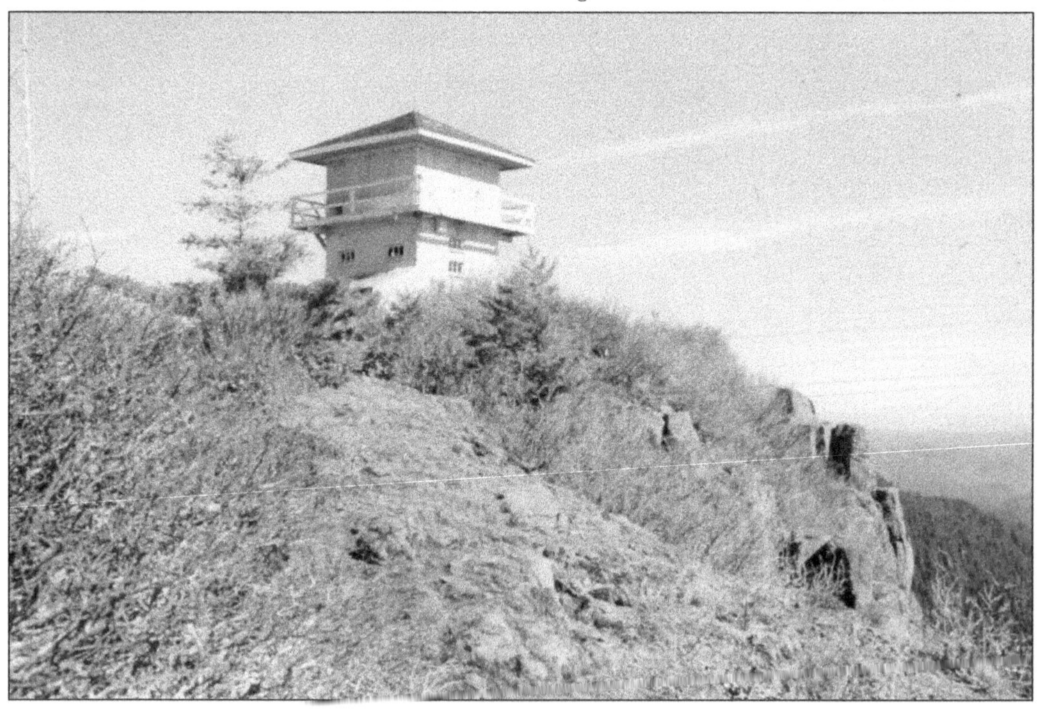

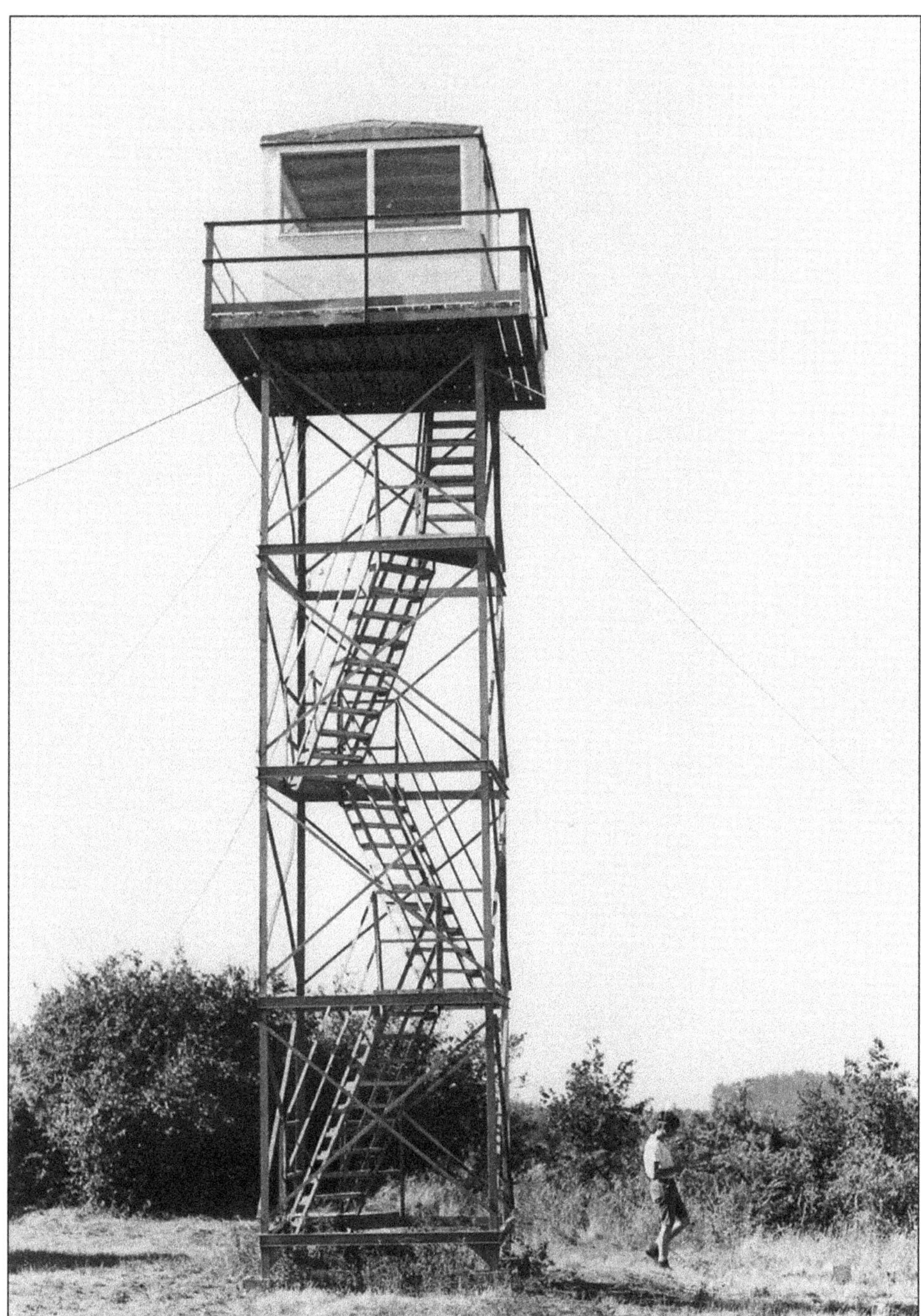

The Western Lane Forest Protective Association (WLFPA) was formed in 1939 to protect forestlands in western Lane County. After a logging accident that broke his neck in three places, in 1950, Harold Morgan took a lookout job at Wolf Point, where the WLFPA had just constructed a new tower. He took his two children and dog Trixie with him. Sometimes, he let the three of them take the canvas water bags to the spring but noted that "by the time they drank and watered the dog several times on the way back, they usually didn't arrive with too much." Morgan worked at Wolf Point for more than 20 years. The original wooden 30-foot-tall tower was burned down to make way for a new steel tower in 1972, pictured. It is now gone. (LCHM.)

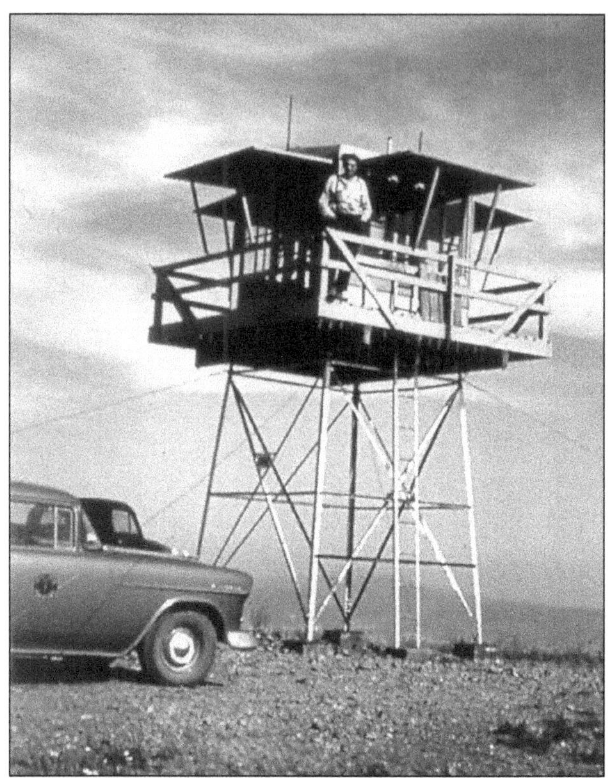

A 30-foot-tall steel lookout tower was built on Gobblers Knob in Tillamook County in 1951. Abandoned in 1965, it has since been removed. This is one of 10 summits in Oregon with the name Gobblers Knob, which is an expression used for a remote feature or a backwater community. (ODF.)

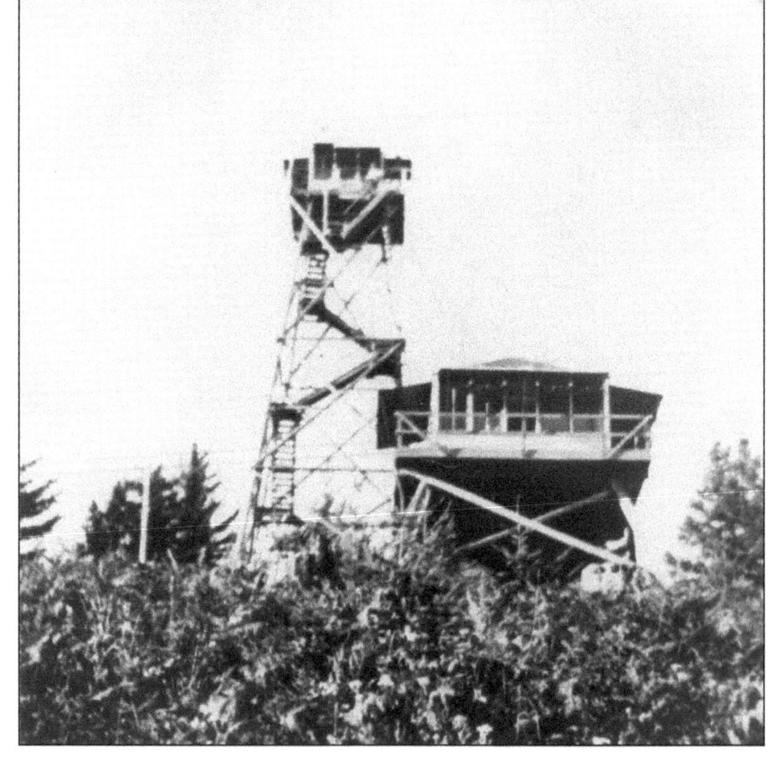

A 15-foot-tall wooden L-4 lookout (foreground) was built on Bell Mountain in 1935. It evidently was not tall enough, because the next year, an 80-foot-tall tower was constructed. The shorter lookout remained in place as sleeping quarters. Both towers were removed in the 1960s. (ODF.)

Pictured is the second lookout on Cannibal Mountain, which was built in 1948. This was the last lookout to be staffed in the Siuslaw National Forest and was last used in 1973. It was torn down and sold for salvage in 1975.

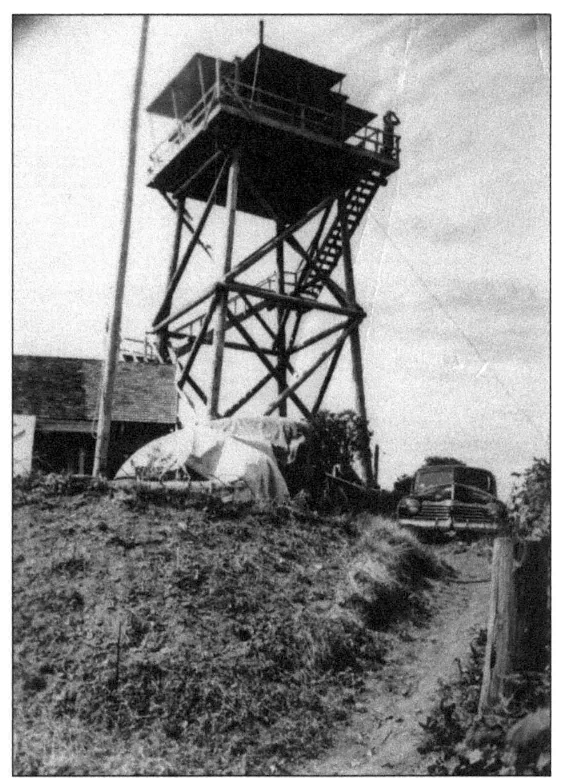

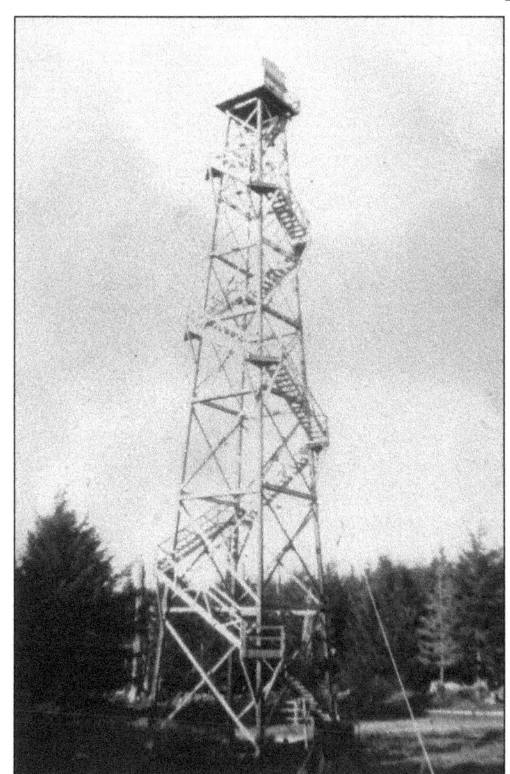

This 1947 photograph shows the 100-foot-tall tower on Little Hebo in Yamhill County. During construction in 1934, a late October storm with high winds hit the area hard. According to the *Oregonian*, Siuslaw National Forest supervisor R.S. Shelley was worried that the tower had been blown over, but it withstood the winds.

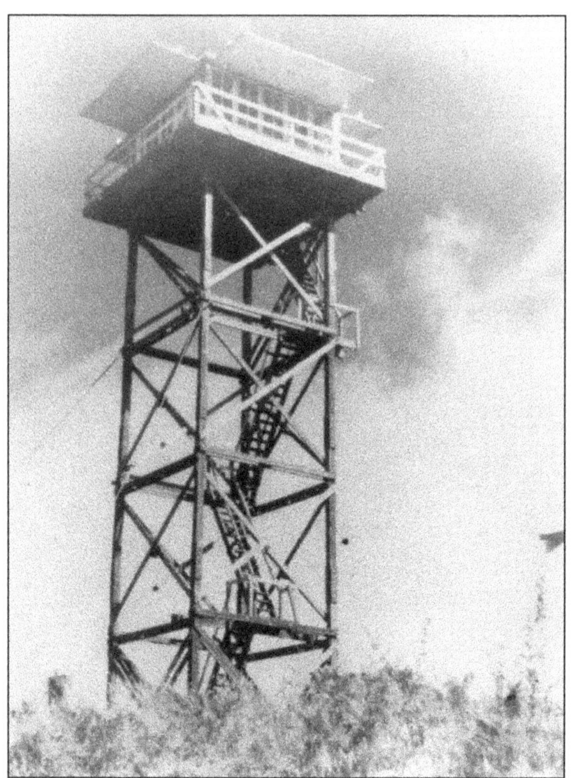

Positioned on exposed mountaintops, lookouts are especially susceptible to the forces of nature. Pictured is the second Buzzard Butte lookout, which was built in 1948 to replace an earlier tower built in the 1930s. The second tower was severely damaged in a storm in 1955 and had to be torn down.

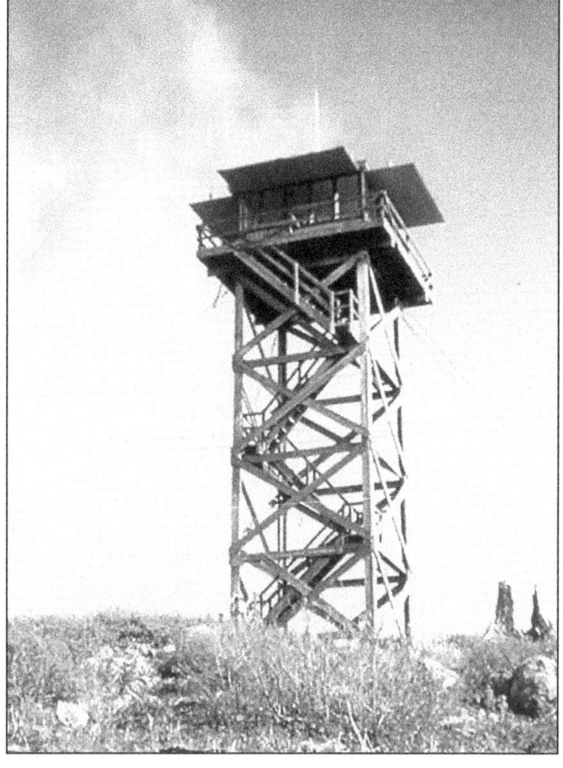

A crow's nest was installed 90 feet up a tree on the south summit of Saddle Mountain in 1932. It was followed by a 40-foot-tall tower (pictured) that was built in 1947. Phyllis Martin staffed the tower during the summer of 1957, between her sophomore and junior years at the University of Oregon. She recalled that the experience "was kind of like going to camp, only all by yourself." (ODF.)

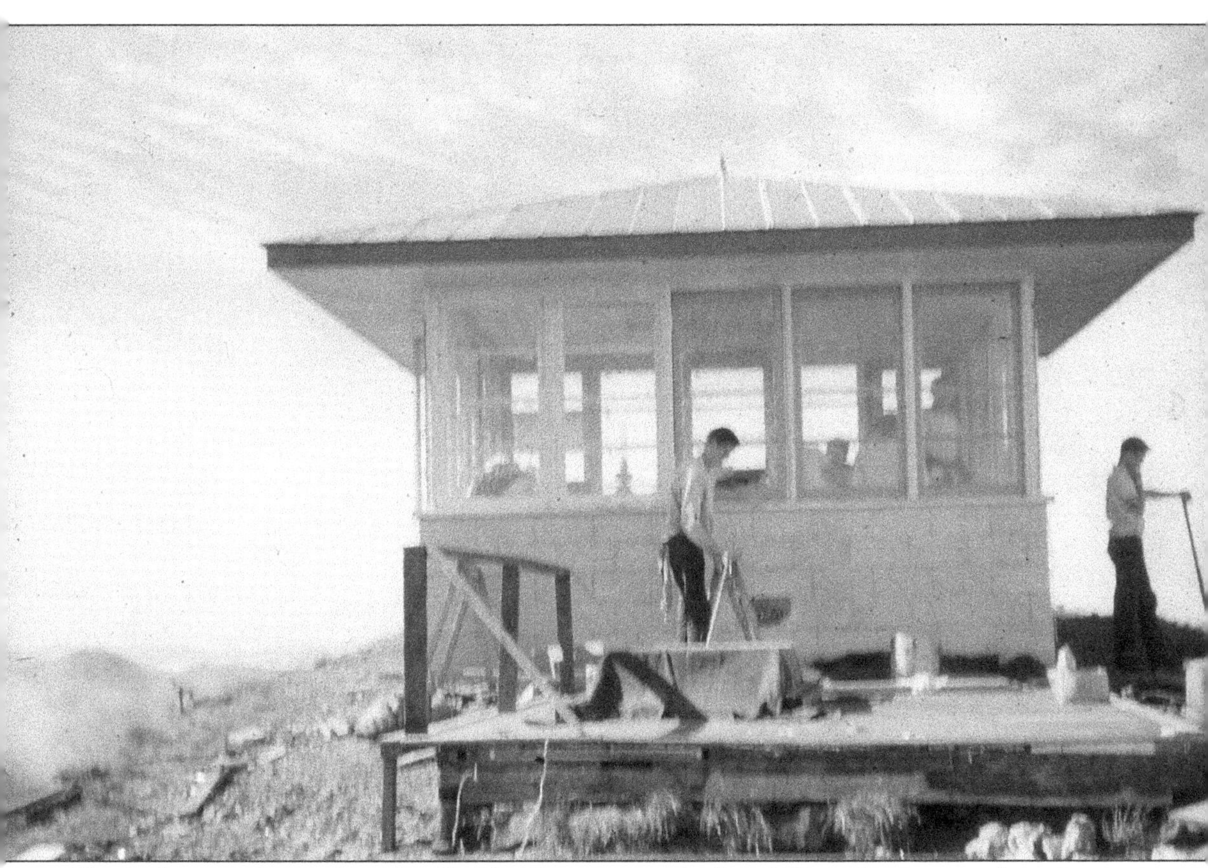

This lookout on the north peak of Saddle Mountain was built in 1953, replacing an earlier cabin built in 1920. More than 21,000 pounds of tools, cement, lumber, and other material needed to be hauled to the summit to construct this lookout, but the trail was too narrow and steep to use pack animals. Helicopter pilot Dean Johnson and his copilot Fred Hill offered to fly the materials to the top, so a small 15-square-foot landing platform was built on the summit, pictured in the foreground. After 85 flights, all the materials were dropped on the summit, and workers spent the next four weeks constructing the new lookout. The total cost came to $4,500, far less than the $10,000 it would have cost if the materials had been packed in by trail. (ODF.)

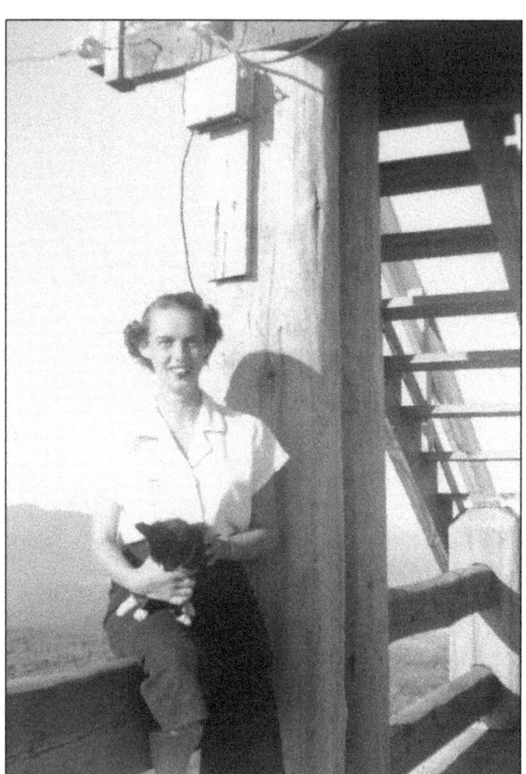

Life on a lookout could be lonely, so pets were sometimes brought along as company. The pet of choice was usually a dog, but this photograph shows Margaret Nutt and her cat at the Marys Peak lookout in 1954.

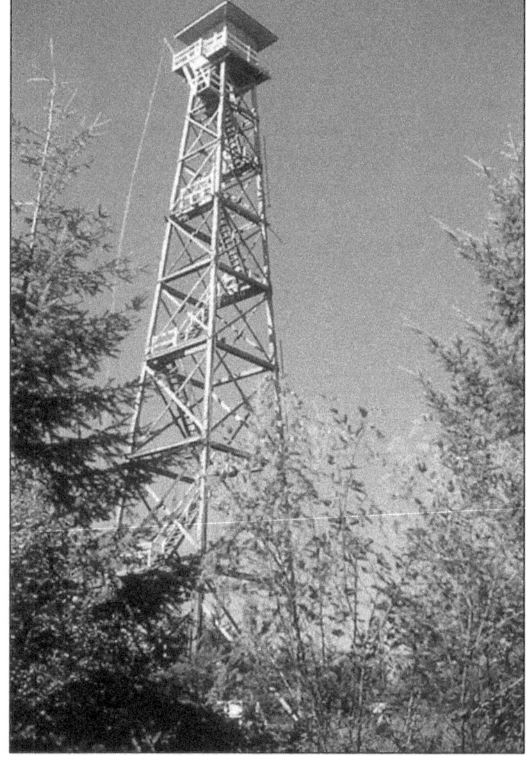

This undated photograph shows the 100-foot-tall tower that was built on Buxton Mountain in 1941. It was replaced by another tower in 1963, which was removed in 1998. The mountain gets its name from a nearby community named for Oregon settler Henry T. Buxton. (ODF.)

Mount Hebo was home to numerous different lookout structures between 1913 and 1978. Pictured is the 67-foot-tall tower that was built in 1950. The photograph shows the tower encased in a thick coating of ice from freezing rain that hit the area in January 1969.

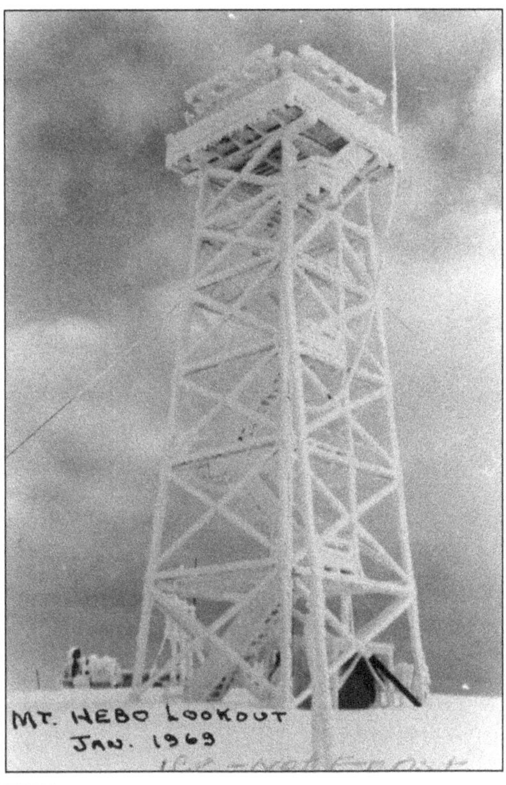

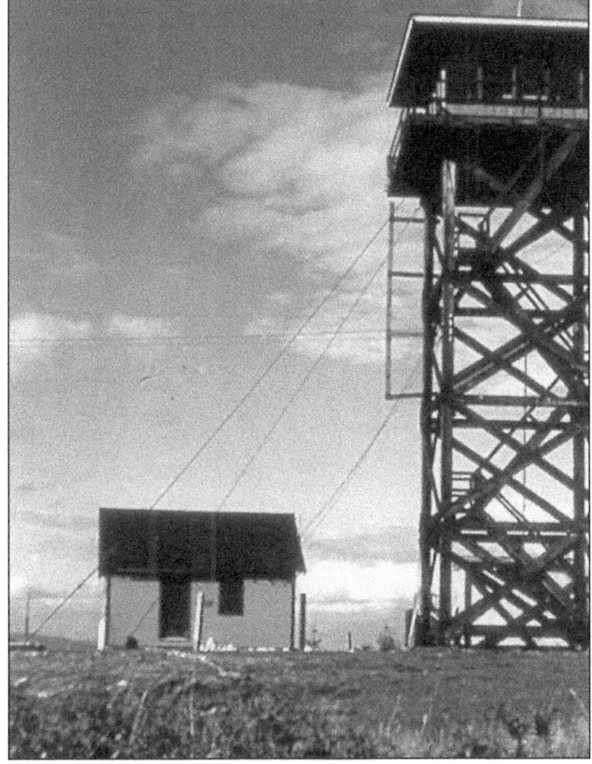

The 40-foot-tall High Heaven lookout was built in 1953. It was staffed for just over a decade before being abandoned in 1965. It suffered from neglect and vandalism, and despite efforts to preserve the tower, it was torn down in 1998. It was the last standing lookout in the North Coast region. (ODF.)

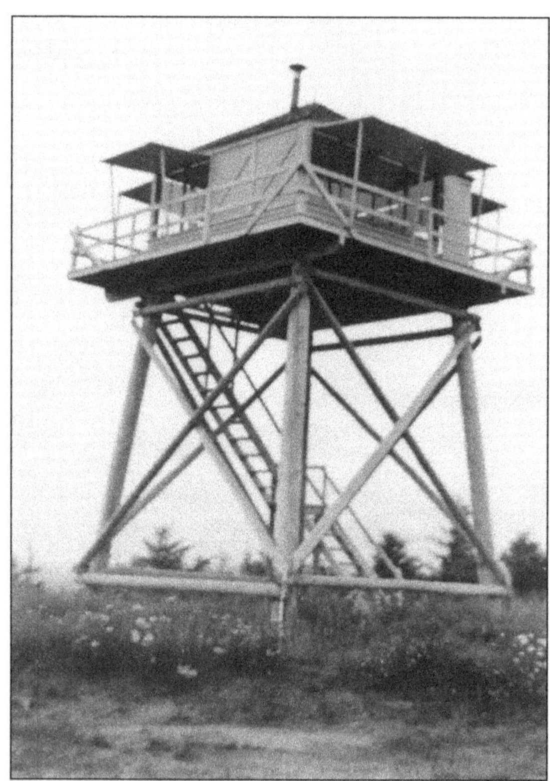

The Siltcoos lookout was built in 1933 and sat within half a mile of the Pacific Ocean at an elevation of just 438 feet. Despite the odd location, the lookout was used for more than 20 years. During World War II, the tower guard watched not only for wildfire smoke but for submarines. The tower is long gone, and the site now sits within the Oregon Dunes National Recreation Area.

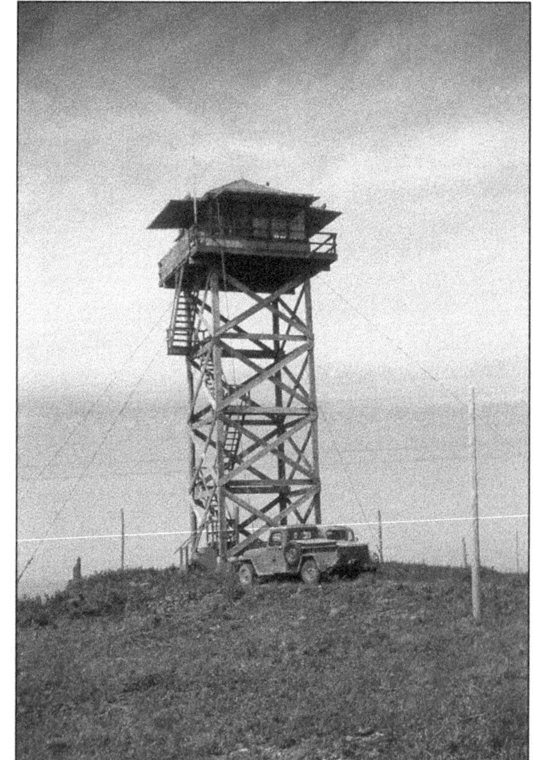

Nicolai Mountain has been a lookout site since a crow's nest was established there in 1939. The pictured tower was built in 1951. After being abandoned for 30 years, it was finally removed in 1994. (ODF.)

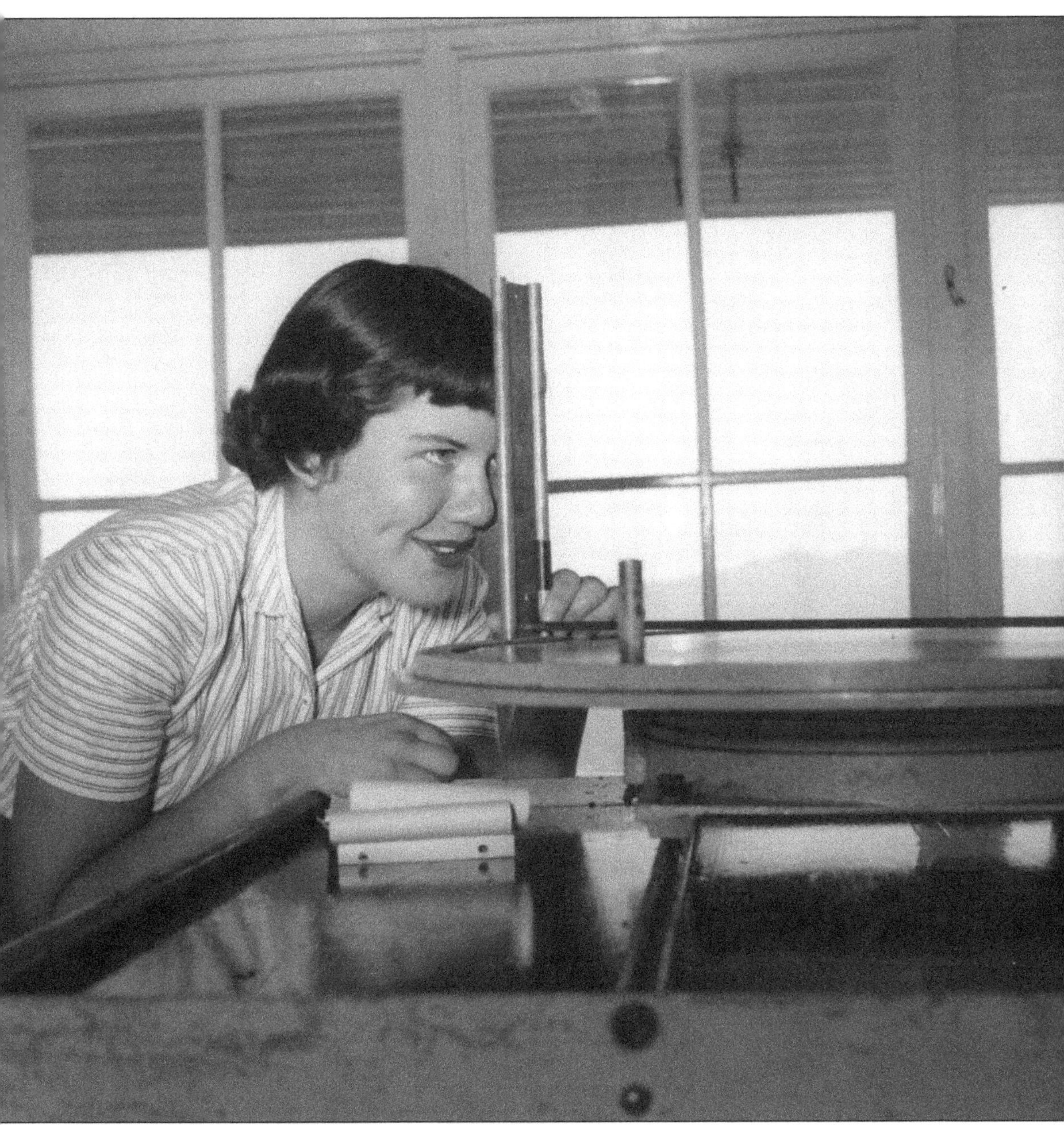

In 1944, six female University of Oregon students worked as lookouts in the northern Coast Range, an area that was heavily burned during a series of four fires between 1933 and 1951 that were collectively known as the Tillamook Burn. These students became known as the "cloud girls" and were the first of many female lookouts to work for the Oregon Department of Forestry. Treva Harden (pictured) was one of these girls, and spent several summers in the 1950s working at the Round Top lookout. For companionship, she brought her dog Freckles. She had to chop her own wood, carry water from the spring 100 yards down the mountain, and use oil lamps for nighttime illumination. But despite the rustic conditions, she loved lookout life, telling a newspaper reporter, "Life up here is wonderful. It's so peaceful." (Forest History Society.)

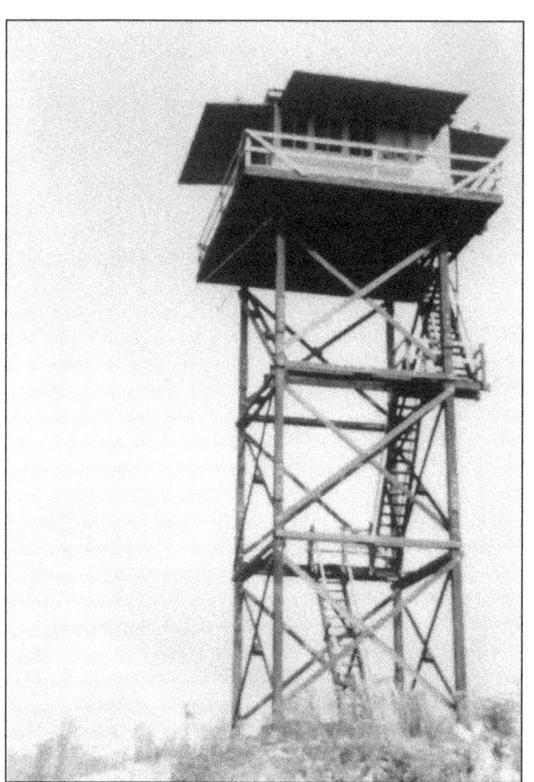

This photograph shows the 40-foot-tall lookout on Blodgett Peak one year after it was built in 1943. The tower stood here for only 10 years. The Forest Service spent $5,200 in 1953 to move the structure to Cummins Peak.

This L-4 cabin was built on Cougar Mountain in the 1940s. Lookouts never knew what unusual task they might have to undertake. Irvin Hussey, the lookout at Cougar Mountain in 1943, got a surprise when Eliga Woody showed up. Woody said he had just shot his partner Charles Phillips nearby in self defense. Hussey got on the radio and called the sheriff to come deal with the situation. There was no report on the fates of Woody and Phillips.

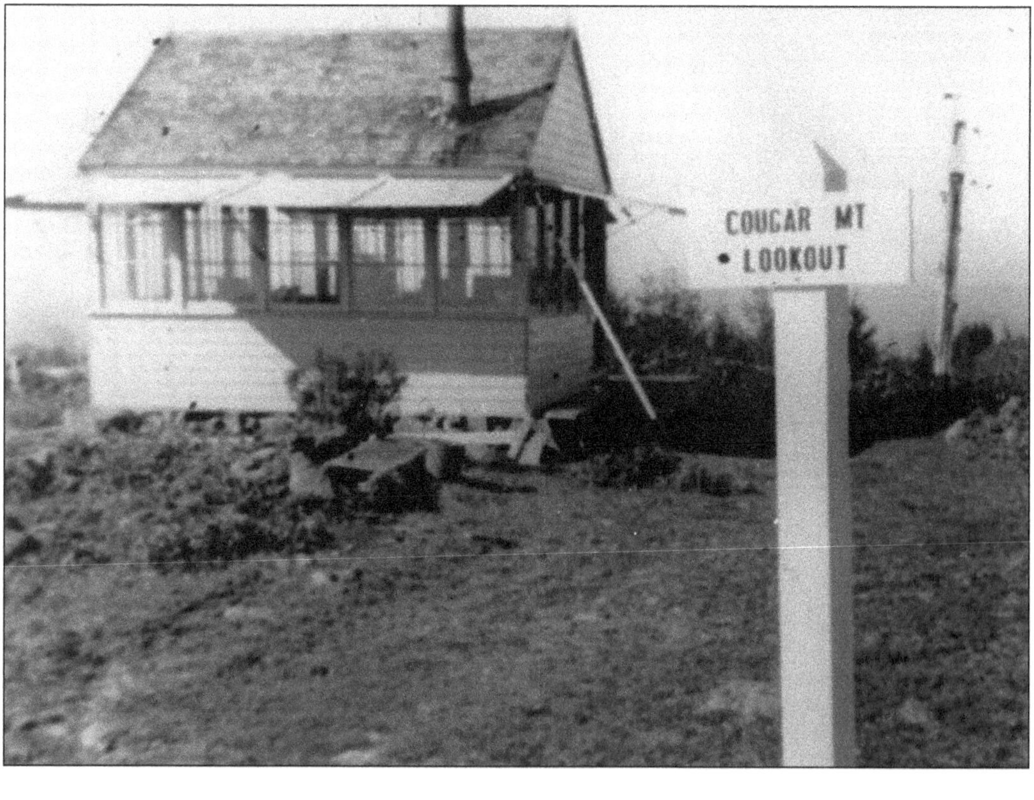

Two

North Cascades

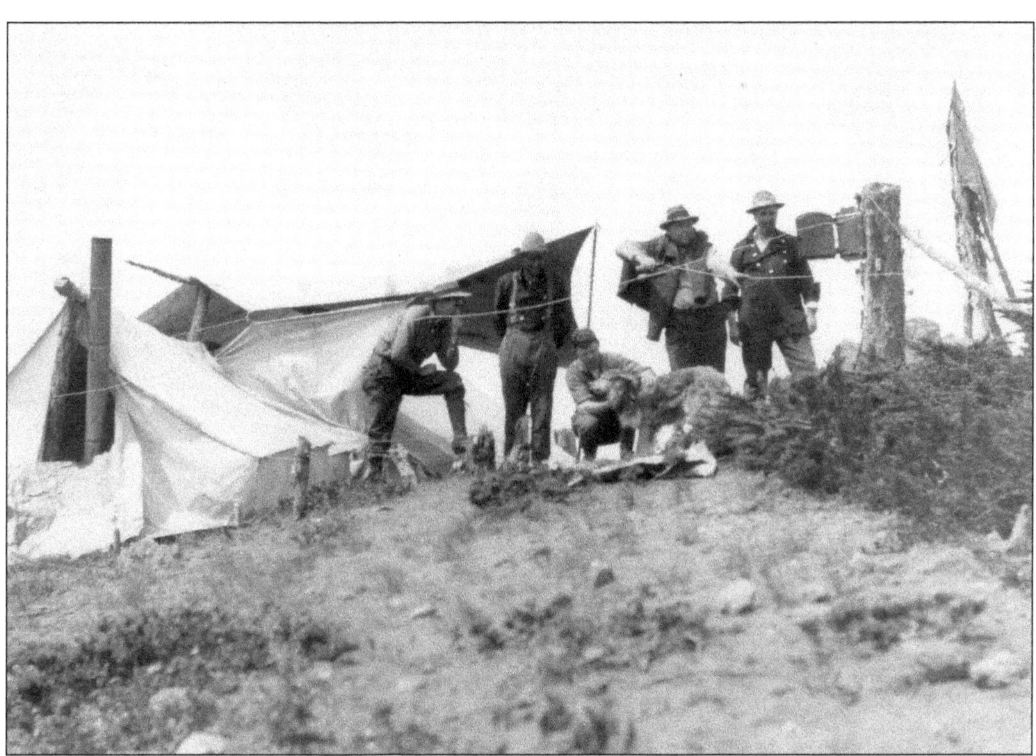

This 1912 photograph shows a group of men at the primitive lookout camp that was established on Battle Ax Mountain that year. The tree stump on the right sports a telephone box, the only way to communicate with the nearest lookout or ranger station. A cupola cabin was built here in 1925, and then an L-4 tower was constructed in 1951. It was removed in 1969.

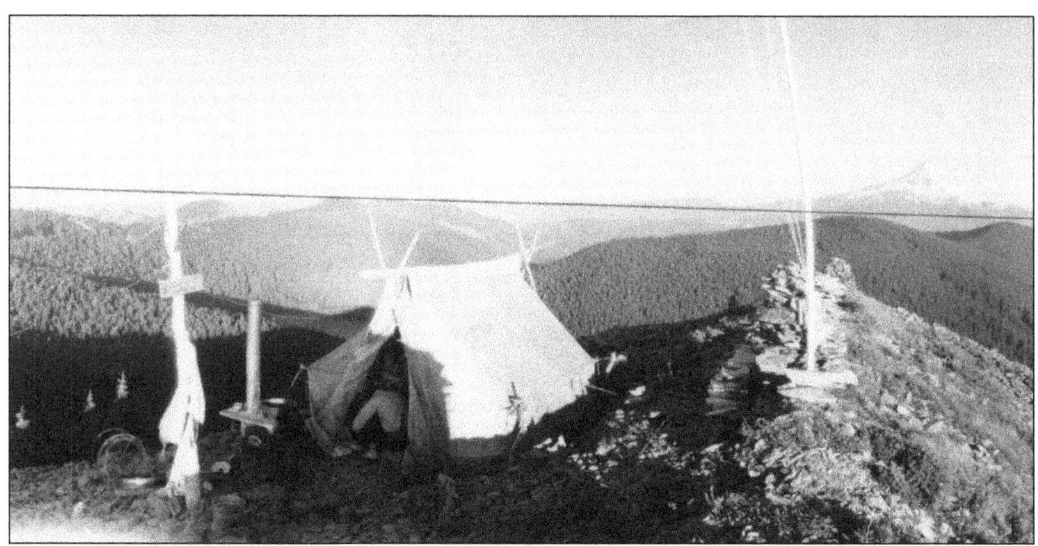

A lookout camp with a fire finder platform was established on Chinidere Mountain in 1929, followed by an L-4 cabin in 1934, which was removed in the 1940s. This photograph shows the primitive living conditions of a tent camp on a mountaintop, although the view of Mount Hood is unbeatable.

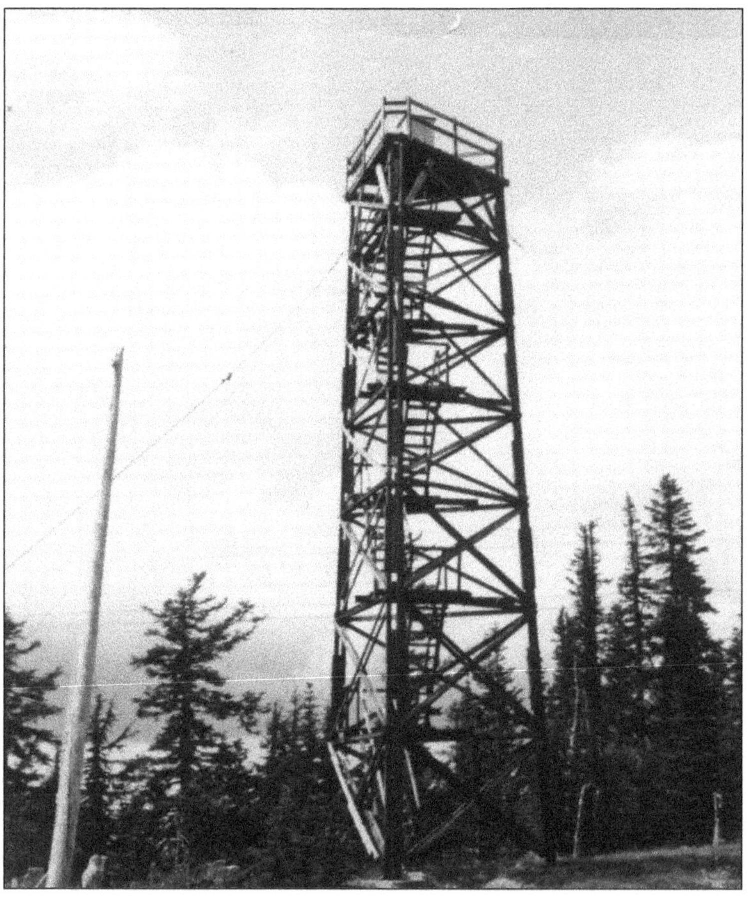

This unusual three-legged platform tower was built on Hawk Mountain in 1919. The tower was abandoned in the 1950s and then removed in the 1960s. The ground cabin that served as living quarters still remains.

Lookout Mountain on the east side of Mount Hood became a fire lookout site in 1911, with an alidade on the summit that was later followed by a cabin with a rooftop viewing platform. Pictured is the final lookout, which was built in 1940. Although it is long gone, the spot is a popular destination for hikers due its up-close views of Mount Hood.

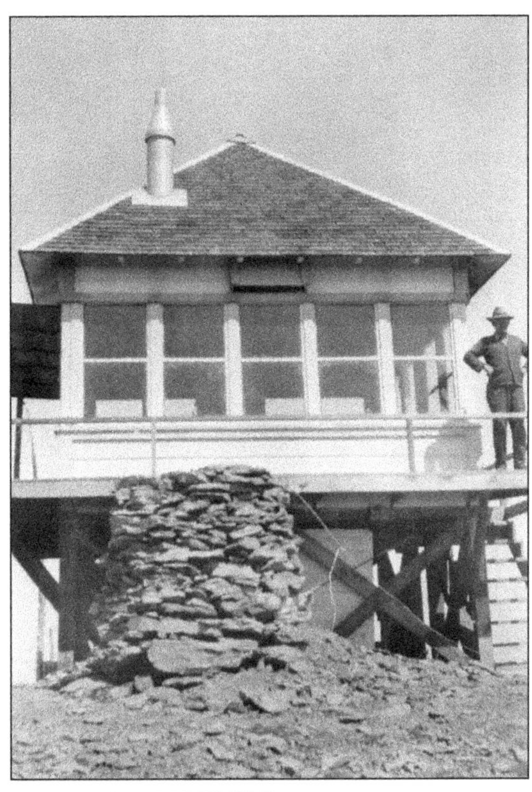

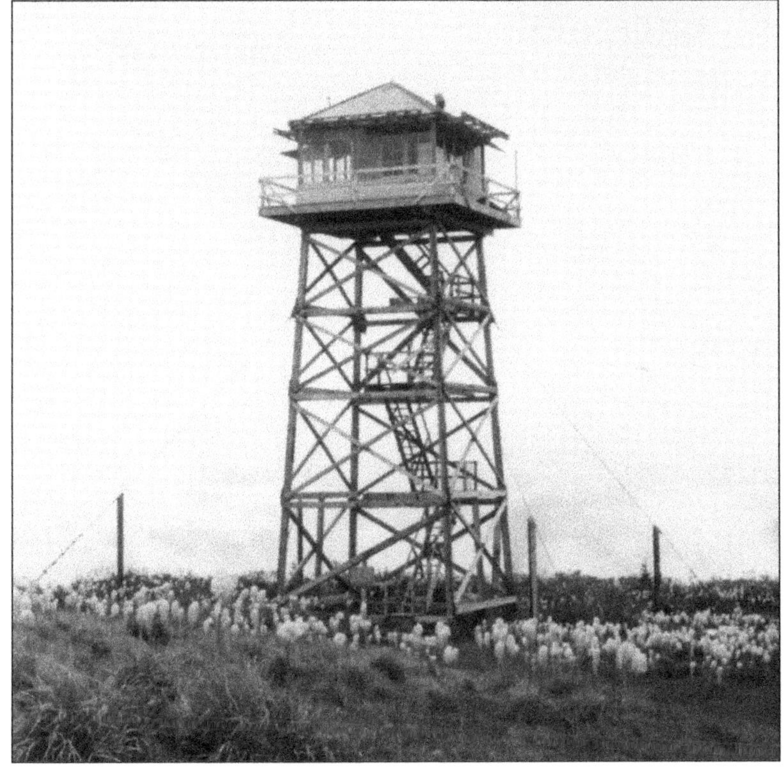

The 41-foot-tall tower on Hickman Butte was built in 1953 and is the second lookout to stand there. It is situated inside the closed Bull Run Watershed on the west side of Mount Hood and is one of only four lookouts still staffed in the Mount Hood National Forest.

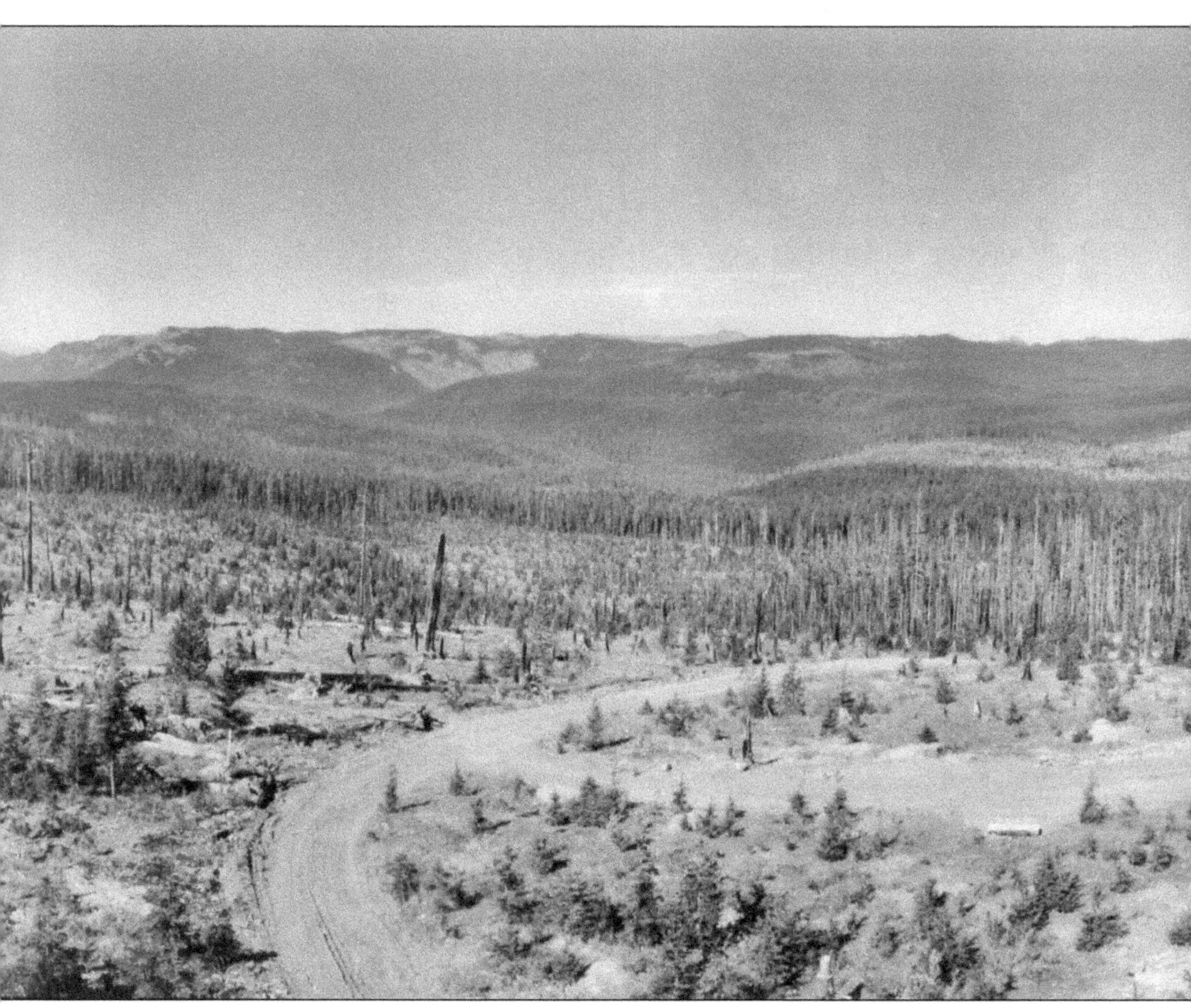

In 1933, the US Forest Service began a project to photograph every forest. Albert Arnst was the project director, and along with six two-man crews, he traveled to lookouts and other high points to take pictures with a special panorama camera invented by William Osborne, the same person who had created the Osborne Fire Finder. Each set of panoramas consisted of three photographs. The project only lasted a few years before funding ran out. This 1933 panorama photograph shows

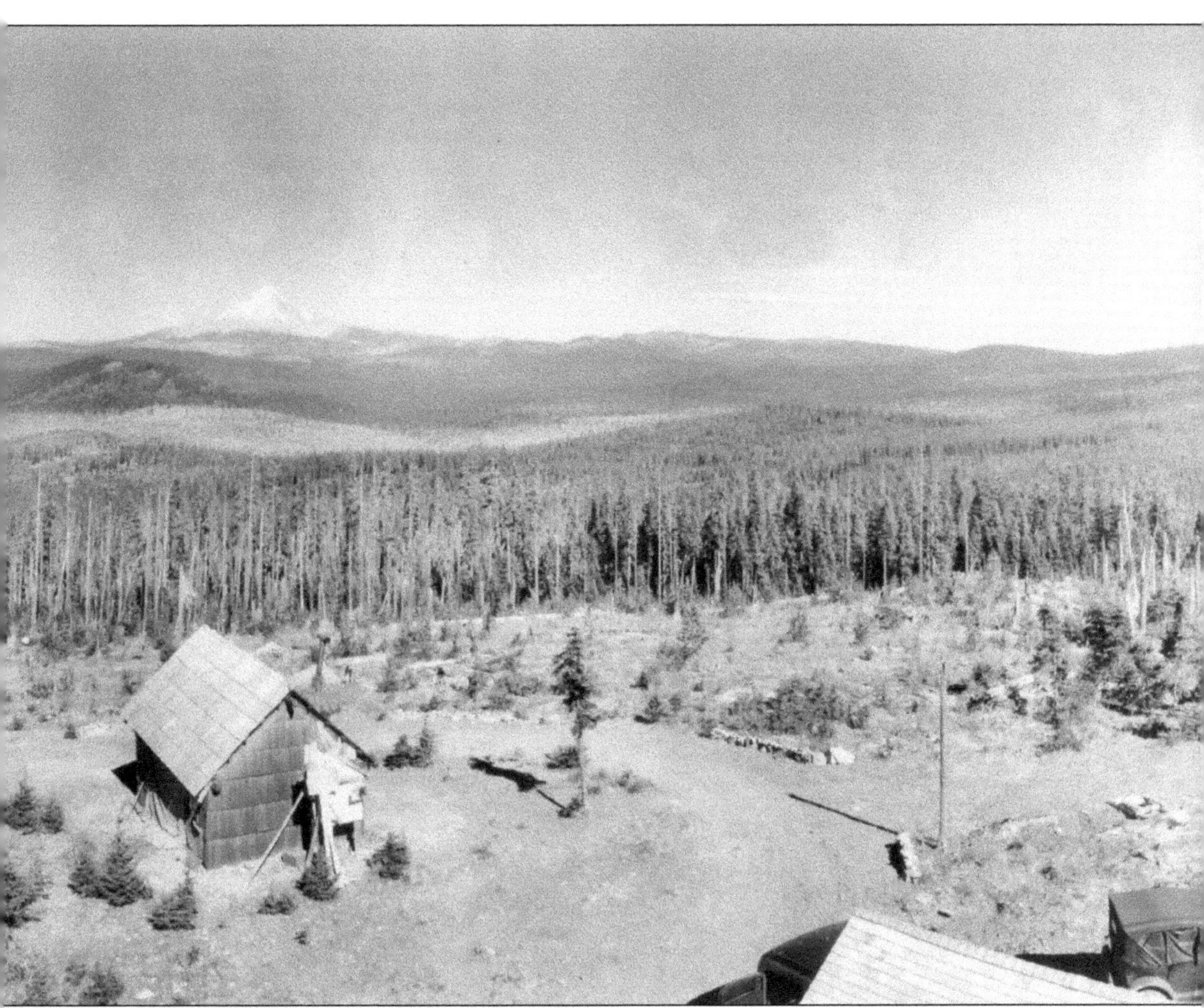

the view from Peavine Mountain looking north. The 26-foot-tall tower there had been finished just that year. David Walker was first stationed here in 1934 and returned the next summer with his new wife, Marcelle; however, their season on the lookout came to an early end when Walker fell off a ladder and dislocated his shoulder.

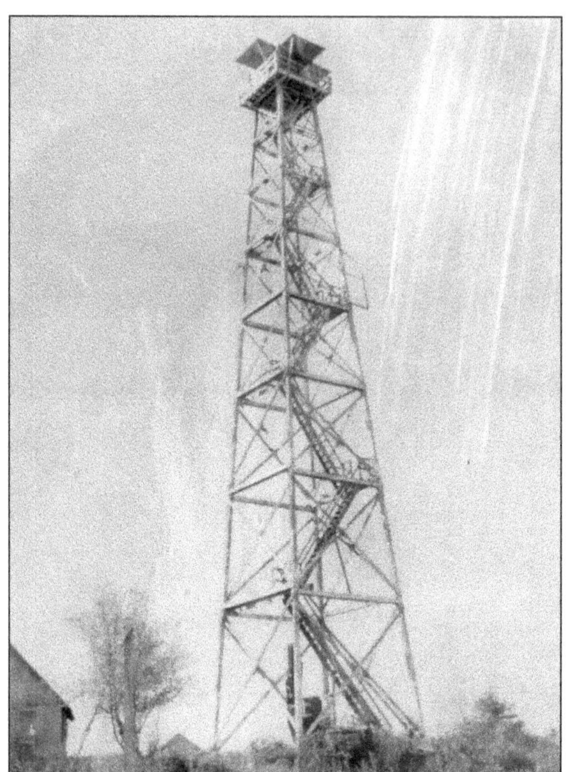

This 100-foot-tall lookout tower was built on Clear Lake Butte in 1932. It was replaced with a 41-foot-tall R-6 lookout in 1962, which still stands today. It is staffed every summer and is available as a rental during the winter.

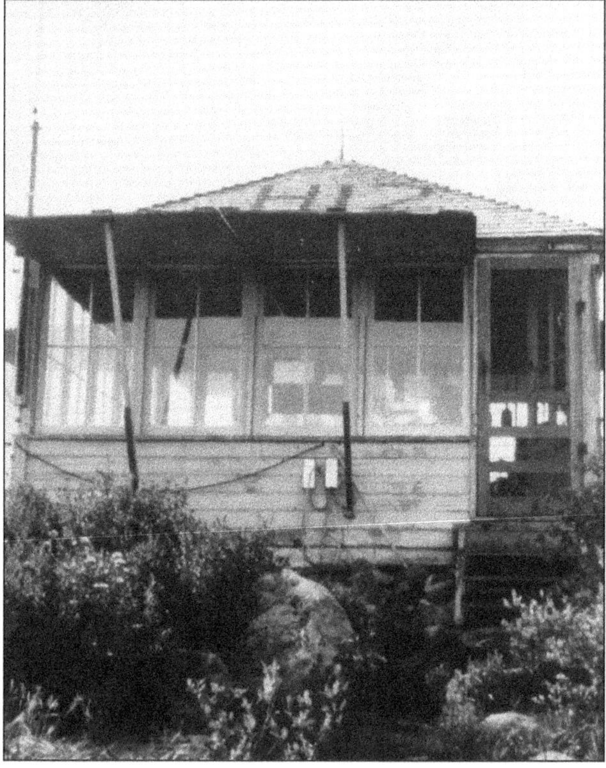

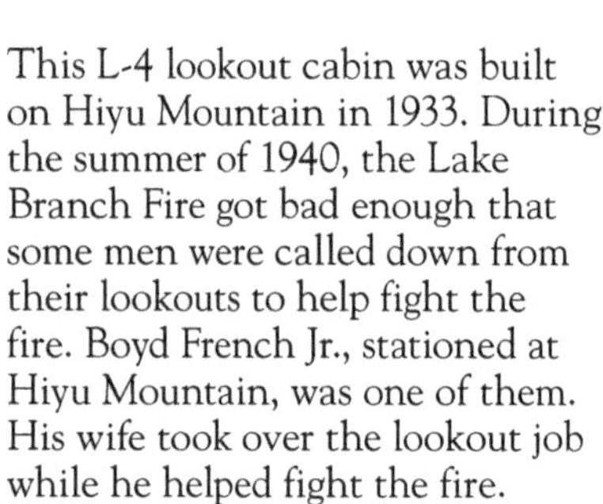

This L-4 lookout cabin was built on Hiyu Mountain in 1933. During the summer of 1940, the Lake Branch Fire got bad enough that some men were called down from their lookouts to help fight the fire. Boyd French Jr., stationed at Hiyu Mountain, was one of them. His wife took over the lookout job while he helped fight the fire.

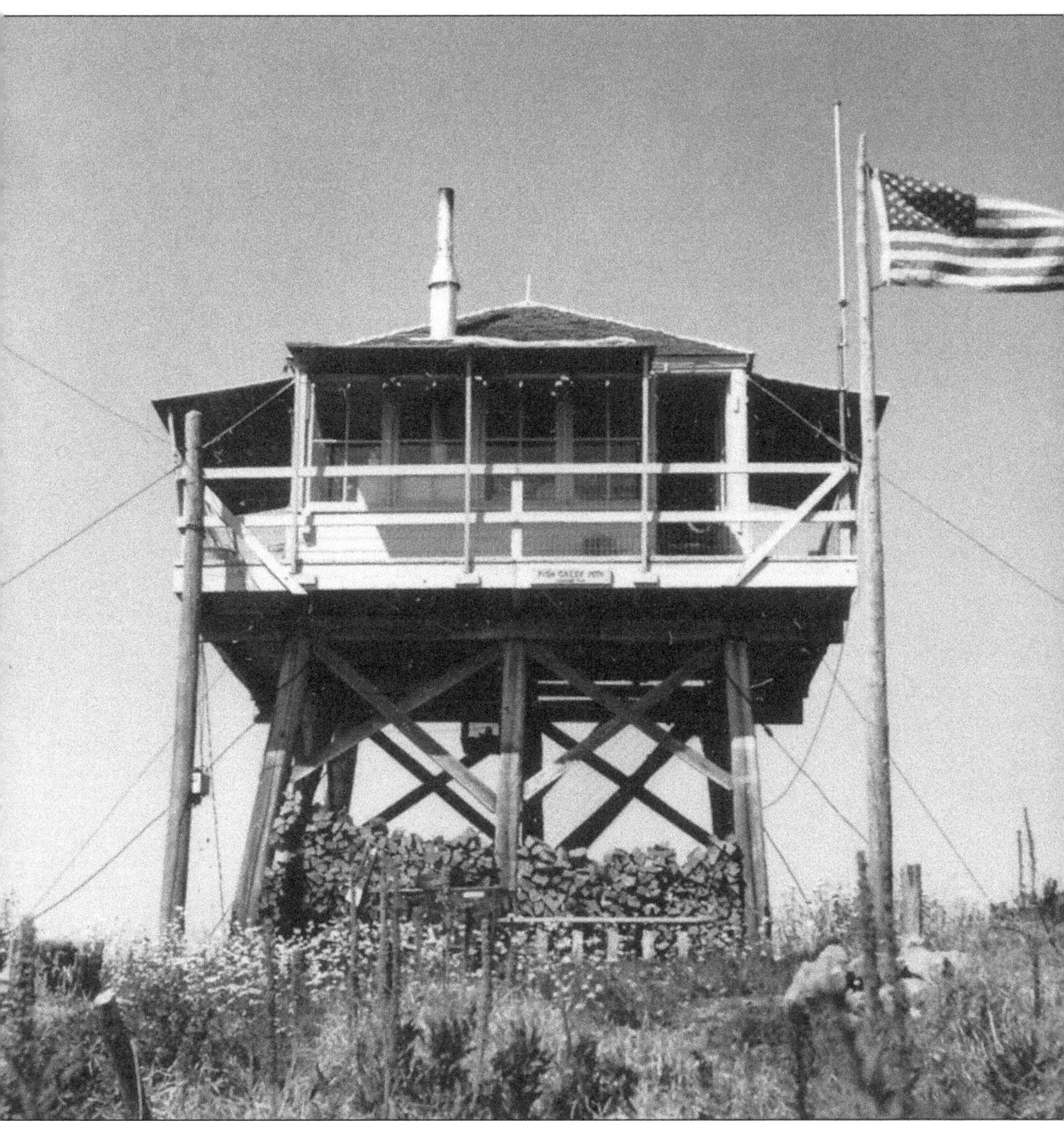

Fish Creek Mountain was the site of a lookout camp starting in 1925. A 12-foot-tall L-4 tower was built in 1933. When Elbert and Helen Brubaker were assigned to work there in 1945, they had to hike eight miles to reach the lookout. Fetching water involved a three-mile round-trip hike down to High Lake, each of them carrying 2.5-gallon water packs. On July 26, a packer brought them groceries and mail. Helen said it was as good as Christmas, especially since Jim Coffman, the grocer, had added five Hershey chocolate bars to the delivery. Despite the rustic conditions and the isolation, Helen wrote in her journal just before leaving the lookout in September: "We will miss the place 'n very often think of our precious moments together 'n how happy we were when we found we were in the family way!" (National Archives.)

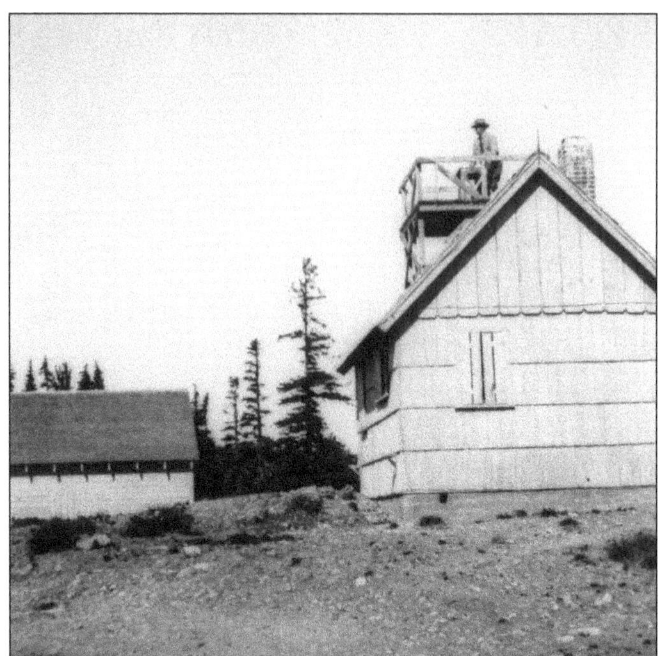

The Red Hill Guard Station was built in the 1930s, and a lookout platform was constructed atop the frame cabin. The station was not actually on Red Hill but about a mile away near a tiny pond known as Perry Lake. The station was demolished sometime after the 1960s, and all that remains are the foundations.

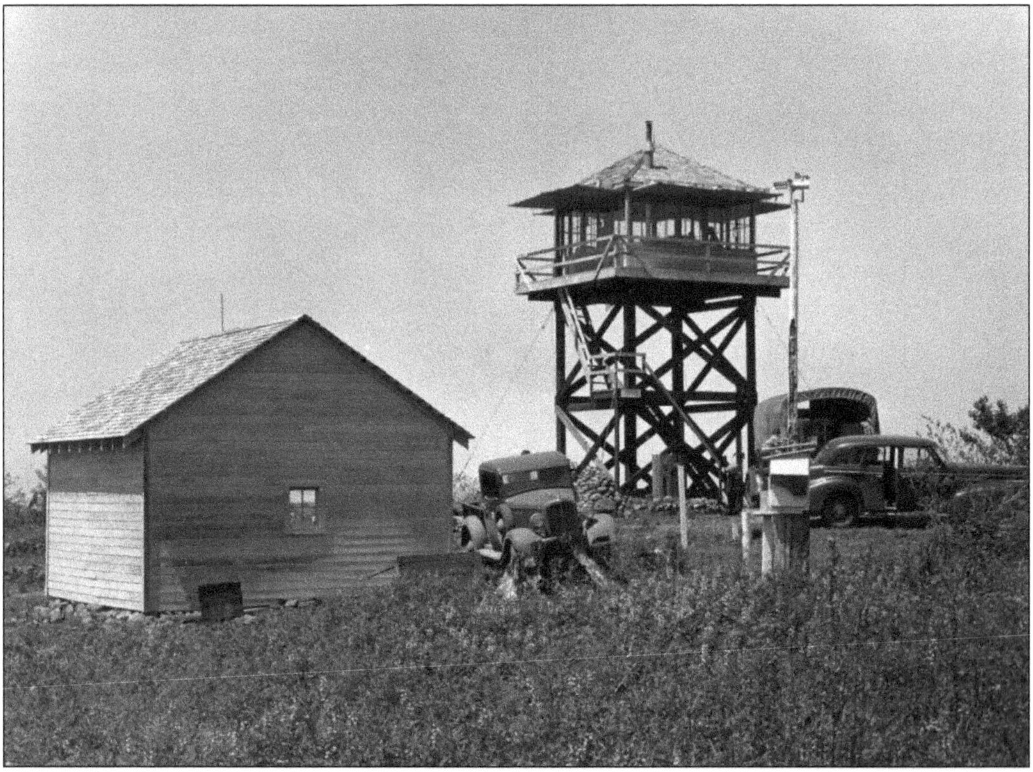

This undated photograph shows the lookout on top of Bedford Point. During the summer of 1943, Wallace Ludgate and his wife signed up for the volunteer forest reserves and staffed the lookout for a week. The reserves were necessary since the war created a labor shortage. Many people used their vacation time to be a reserve volunteer.

A cupola cabin was built on Indian Mountain in 1916. A reporter from the *Oregonian* visited Indian Mountain in 1931 and spoke to lookout Luther Elliff and his wife. Mrs. Elliff reported that she found plenty to keep her busy while her husband was working, including listening to the radio, picking huckleberries, crocheting, and entertaining visitors who hiked up from Wahtum Lake.

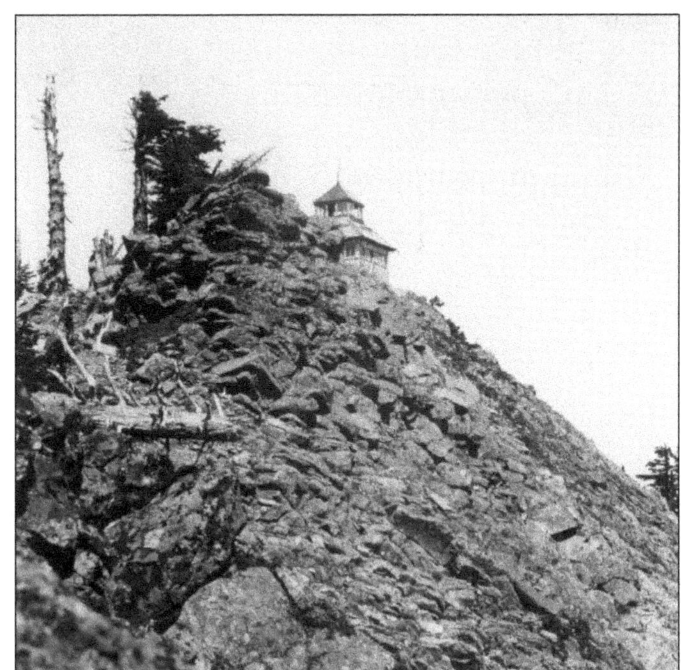

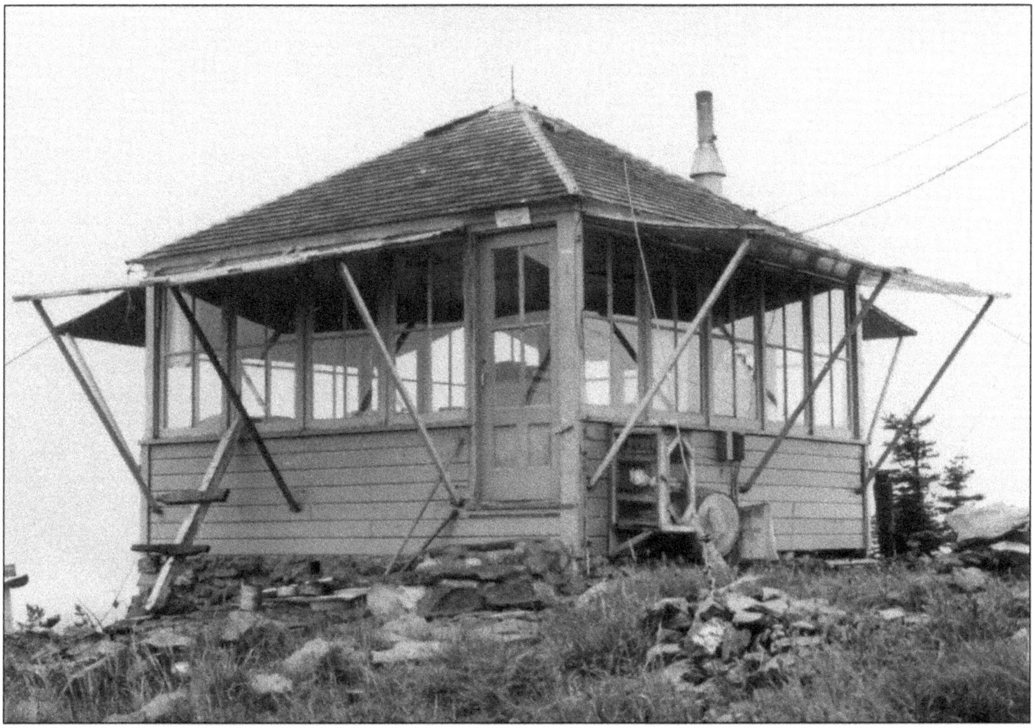

A lookout camp was established on Bonney Butte in 1925, followed by this L-4 cabin in 1933. It ceased to be used in 1964 and was burned to the ground in 1967. Because of its unique position along a bird flight path, Bonney Butte is now the site of an annual raptor survey undertaken by Hawkwatch International. Trained observers count the number of migrating owls, eagles, hawks, falcons, and other birds of prey that they see.

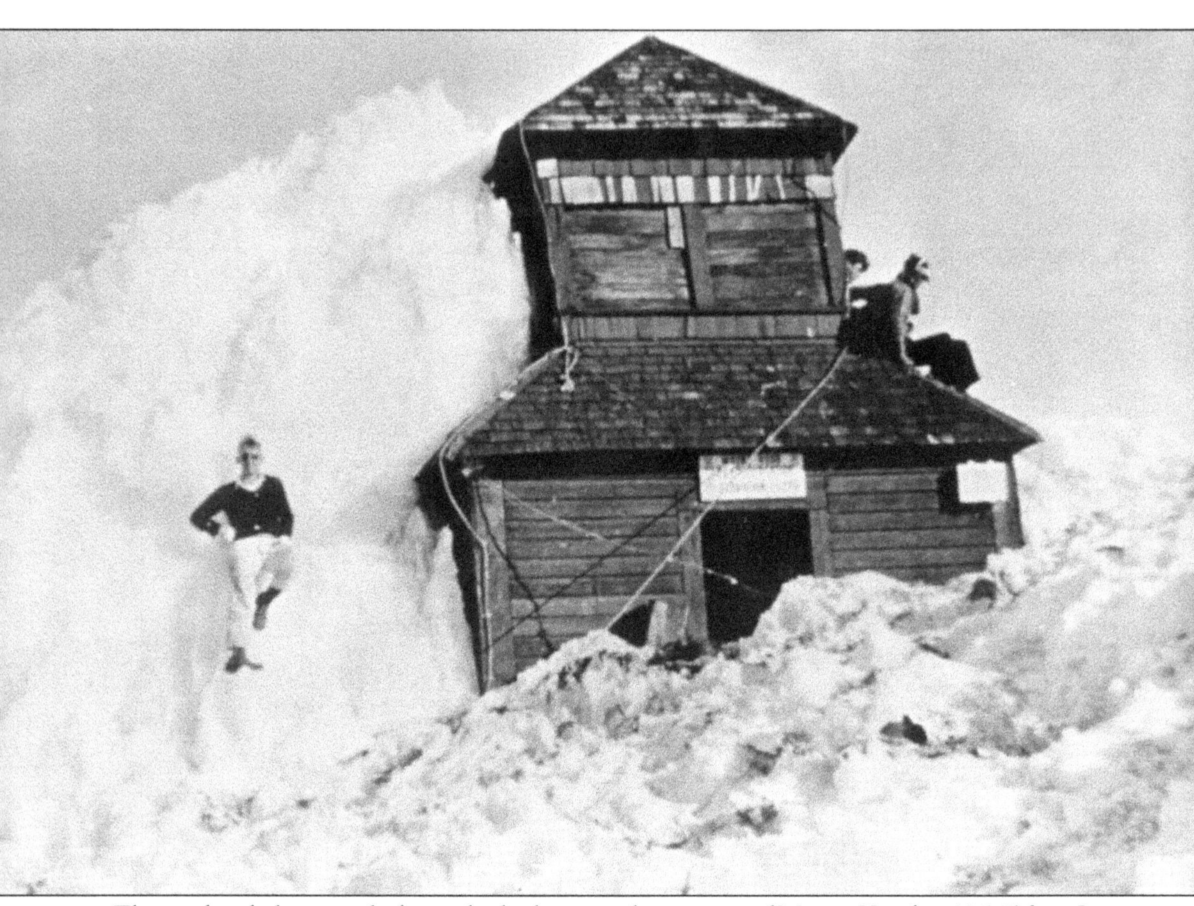

This undated photograph shows the lookout on the summit of Mount Hood at 11,245 feet. It was constructed in 1915 and was the highest lookout ever built in Oregon. During the summer of 1928, Elva Taylor married Robert McVicar—who was working at the lookout—in a mountaintop ceremony. She spent four summers with her husband at the lookout. In 1931, she wrote about her experience for the *Oregonian*: "During the first summer, I heard the following comment at least a hundred times: 'You must think an awful lot of Mr. McVicar to have climbed all the way up here to be married.' " Because of the difficulty of cooking at such a high elevation, she and her husband mostly ate dried fruit, canned goods, ham, hotcakes, and cookies. "How tired we become of cookies!" she wrote. After it was abandoned in the 1930s, the Mount Hood lookout fell off the summit in 1941.

Zigzag Mountain once boasted two separate lookouts. Forest Service employee George Henderson appears in this photograph of the L-4 lookout that was built on West Zigzag in the 1930s. The lookout on East Zigzag, about three miles away, was built around the same time. Both lookouts were removed in the 1960s. (George M. Henderson.)

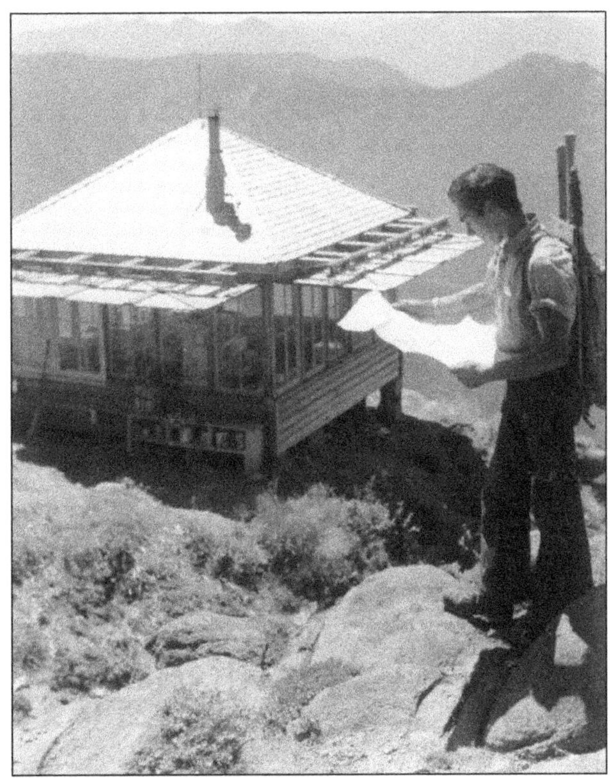

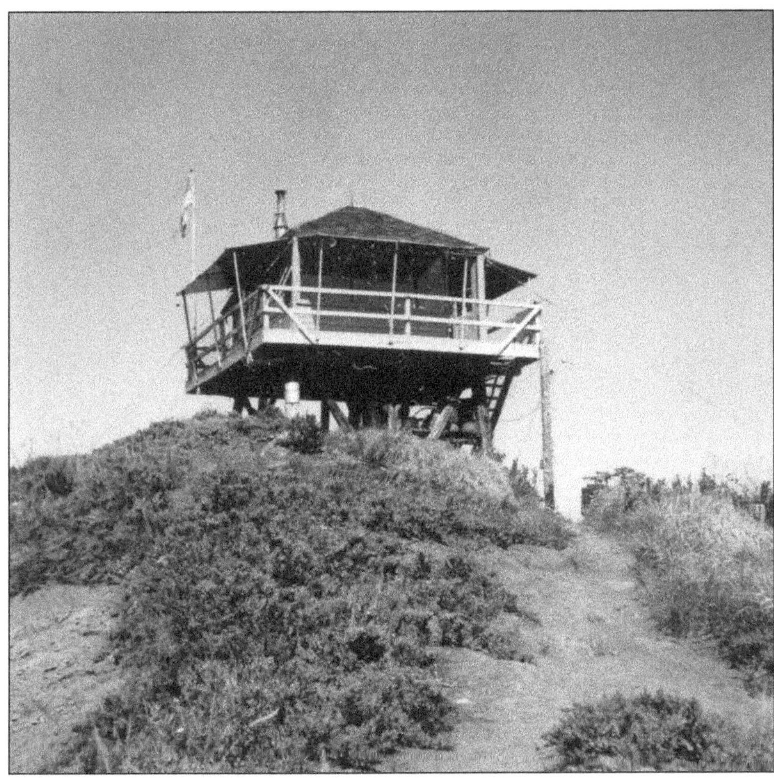

This L-4 lookout was built on Thunder Mountain in 1933. By 1963 the tower was in disrepair, with cracked foundation blocks, rotting beams, and an infestation of carpenter ants. Rather than repair or replace it, the Forest Service removed it. (National Archives.)

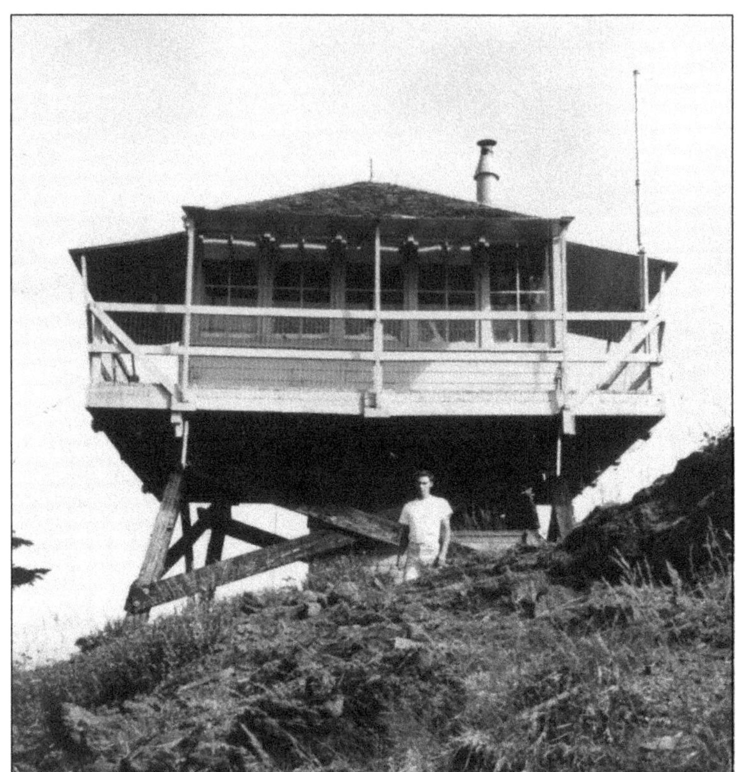

Pictured is the third lookout to stand on Devil's Peak, replacing cabins that were built in 1924 and 1933. This tower was constructed in 1952 but ceased to be staffed in the 1960s. It managed to escape the fate of many deactivated lookouts and has remained standing. It now looks out over the Salmon-Huckleberry Wilderness, which was created in 1984.

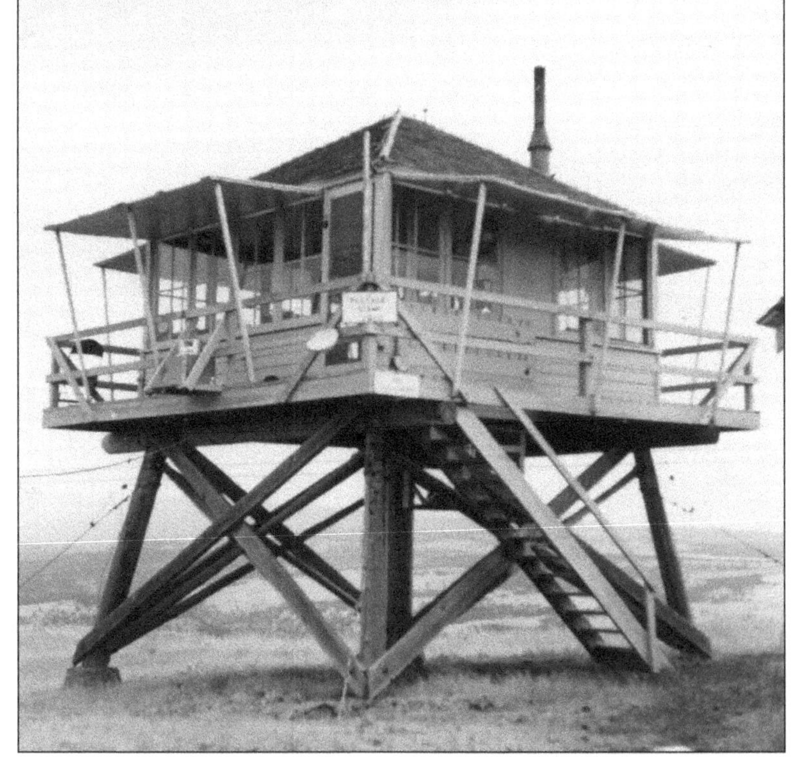

The first lookout on Postage Stamp Butte was built by the Civilian Conservation Corps in 1937. Pictured in 1963 is the second tower, which was built in 1945. It has since been removed. The butte was named in 1925 by James Frankland, who observed that when viewed from a distance, the arrangement of ground colors resembled a letter with a postage stamp in one corner.

A succession of lookouts has stood on Fivemile Butte since the 1920s. Pictured is the fourth and final tower, which was built in 1957. The lookout still stands, and although it is no longer staffed, it is available for the public to rent year-round.

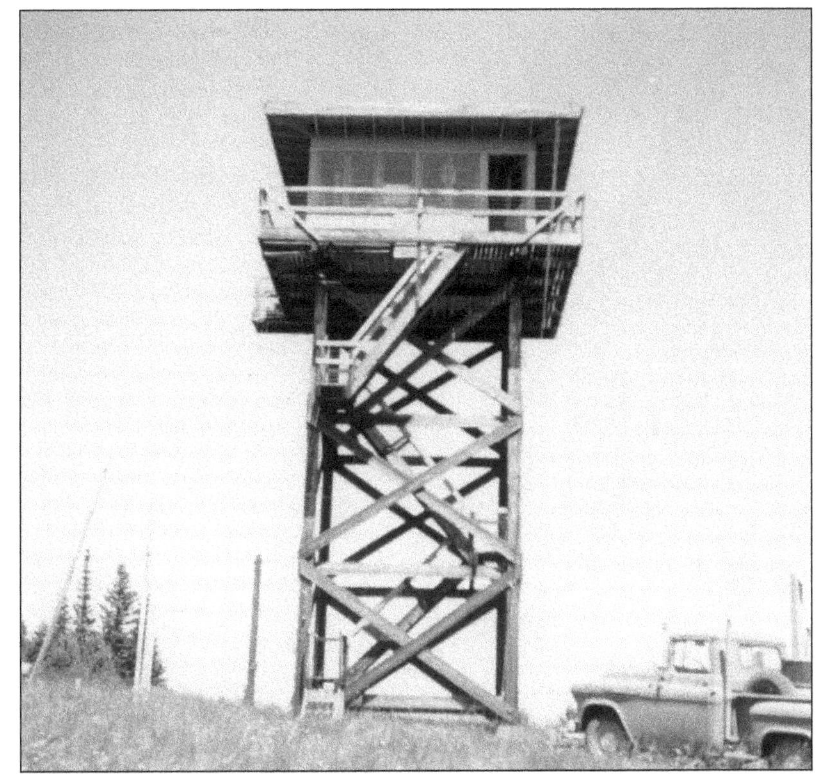

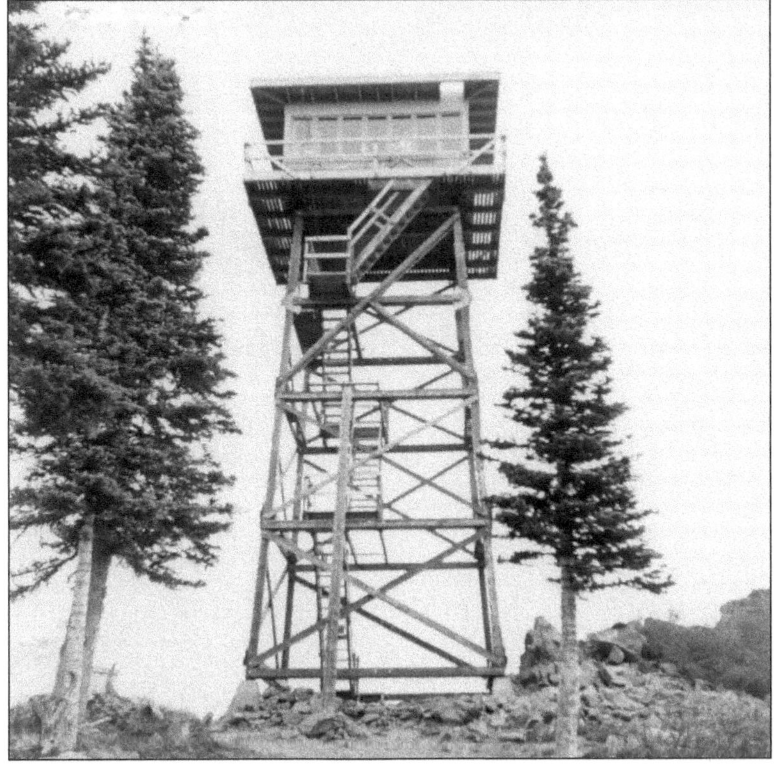

Flag Point has been the site of three different lookouts since 1924. This is the third tower, which was built in 1973. The 41-foot-tall lookout towers over the Badger Creek Wilderness and is staffed in the summer. It is available to rent in winter for those who have the stamina for the 11-mile ski or snowshoe trek.

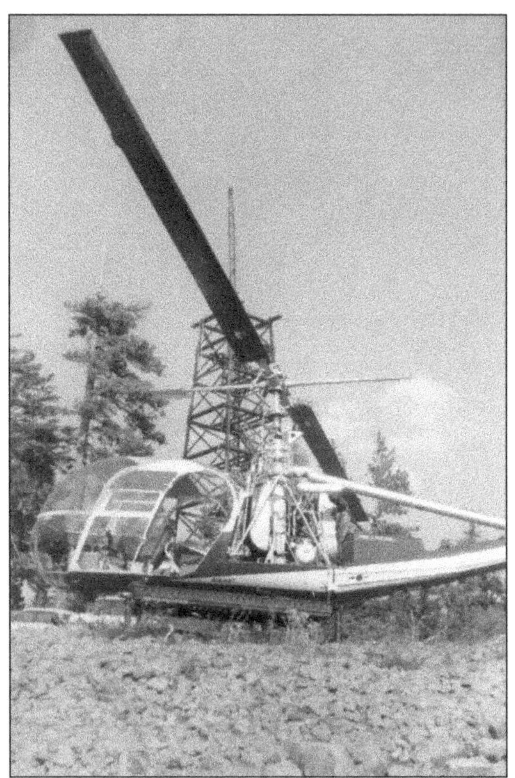

This 1962 photograph shows the construction of the third lookout tower on Post Point. By 1994 the tower was no longer in use, and the Forest Service attempted to move the cab to a different location; however, it was too heavy for the helicopter and was dropped to the ground. Lumber, bolts, and nails can still be seen scattered around the summit.

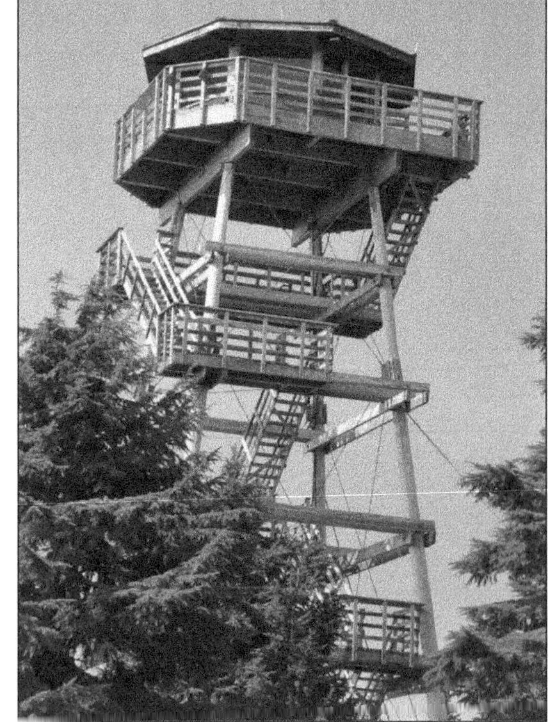

Starting in 1940, Sisi Butte was home to a 41-foot-tall tower. The Forest Service removed it in 1996 and built a new 50-foot-tall tower with an unusual eight-sided cab, pictured in 2014. The tower is still staffed in the summer. (Author's collection.)

Three

Central Cascades

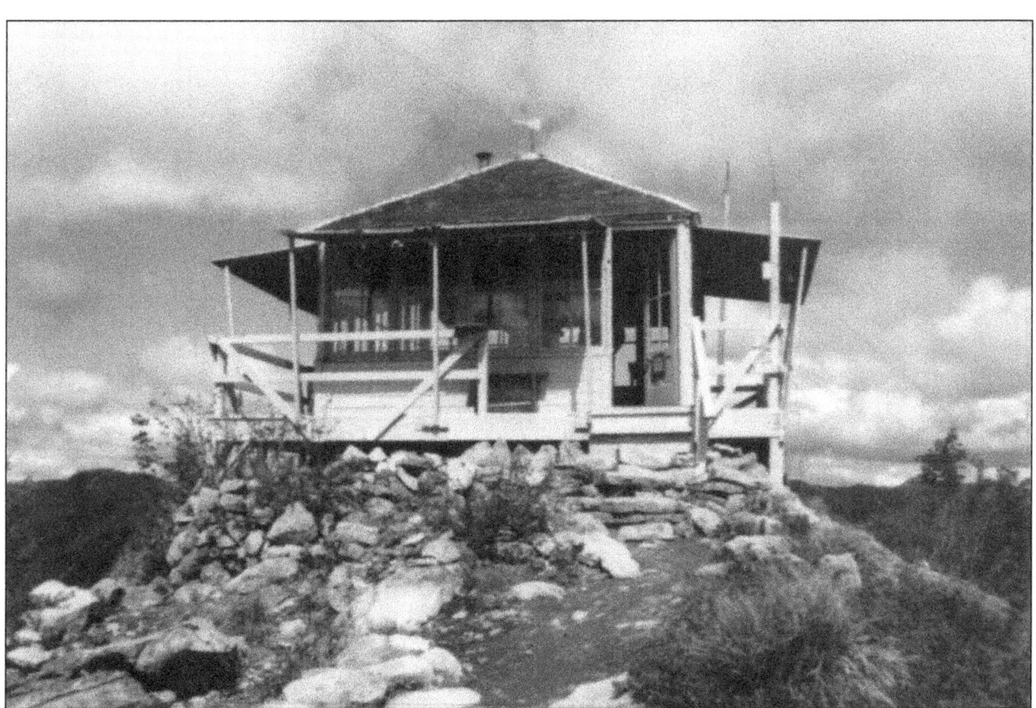

The lookout on Gold Butte was built in 1934 and was last staffed in the 1970s. By 1999, the old building was barely standing, suffering from decades of neglect and vandalism. More than 50 volunteers from the Sand Mountain Society helped restore the lookout between 1999 and 2007, and it is now available to rent.

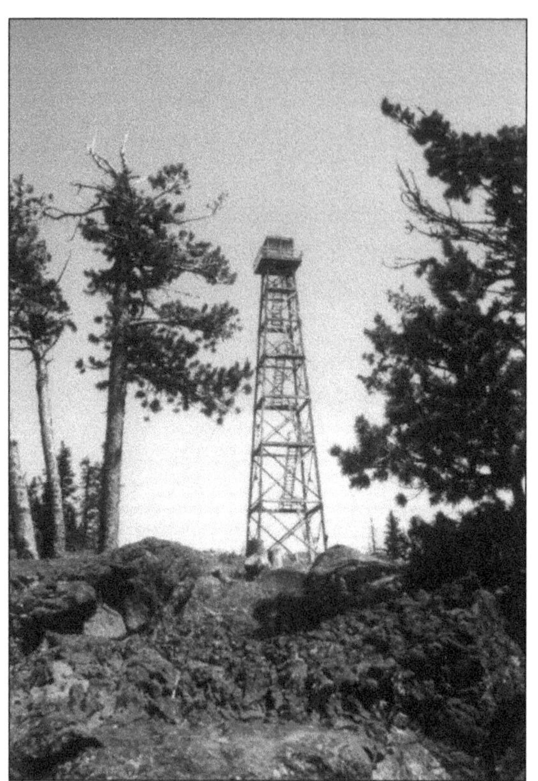

Black Butte has been a lookout site since 1910, when a crow's nest was established on the summit. A succession of lookouts followed, including a cupola cabin in 1923 (which still stands), an 82-foot-tall tower in 1934 (pictured in 1991), and a 65-foot-tall tower in 1995 (which is still staffed every summer).

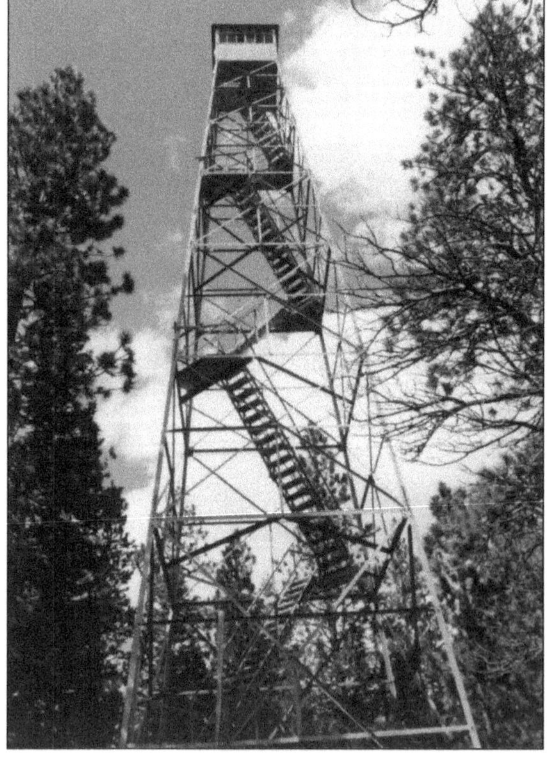

The 86-foot-tall tower on Trout Creek Butte was built in 1933. One summer, the stationed observer saw an aircraft flying by his lookout. He watched it as it flew off toward North Sister, but the tiny silver dot of the airplane disappeared. Thinking it had crashed, the man took a look through his binoculars and saw that it had actually landed. The owner of the plane was eventually fined $50 for operating a motor vehicle in a designated wilderness area.

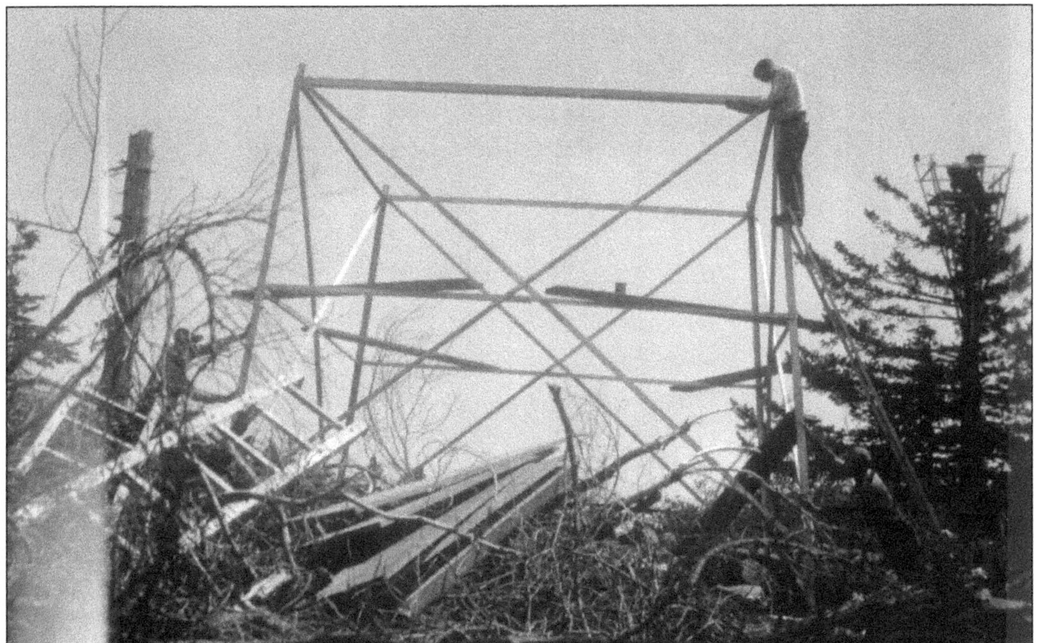

This photograph shows the Saddleblanket Mountain lookout under construction in 1927. The old crow's nest lookout can be seen in the background. The 70-foot-tall tower still stands, although it has been abandoned since the 1960s.

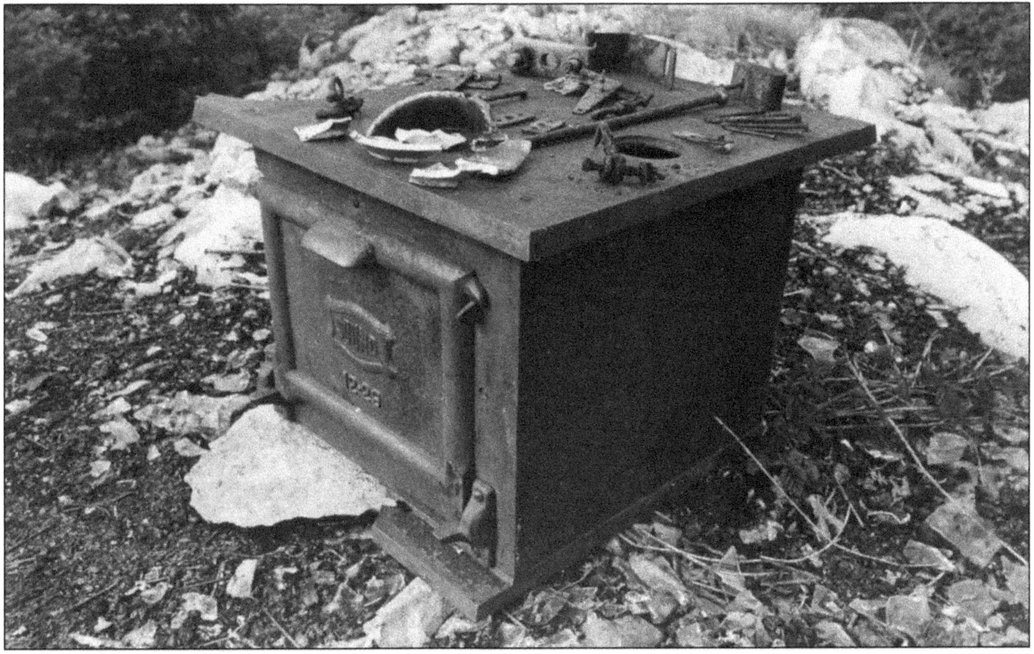

Fred Merten was just 18 in the summer of 1961 when he was stationed at the lookout on Hehe Mountain. Lying awake in the darkness on his first night, he felt the 22-foot-tall tower vibrating. When he mustered the courage to go outside and take a look, he found a doe scratching her back on one of the guy-wires. The lookout was removed a few years later, and this photograph taken around 1968 shows all that was left: the old woodstove. (LCHM.)

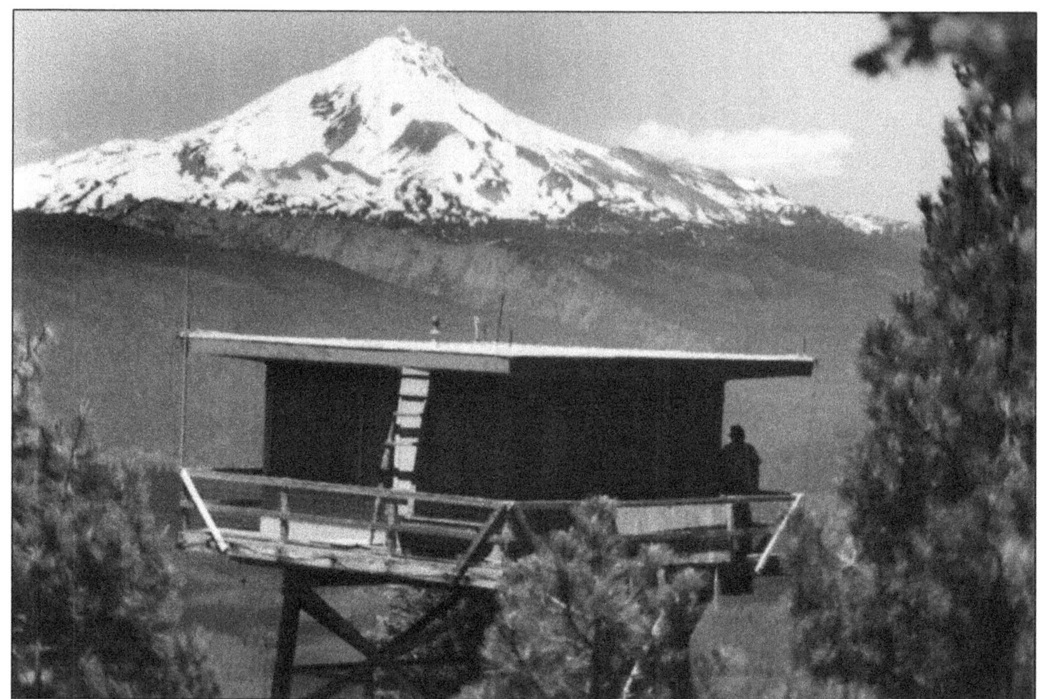

The first lookout on Green Ridge was built in 1933 and later replaced by this 20-foot-tall tower. Rather than being situated atop the ridge, it is on the side, with a view overlooking the valley of the Metolius River. It is staffed in the summer but available as a rental in the spring and fall.

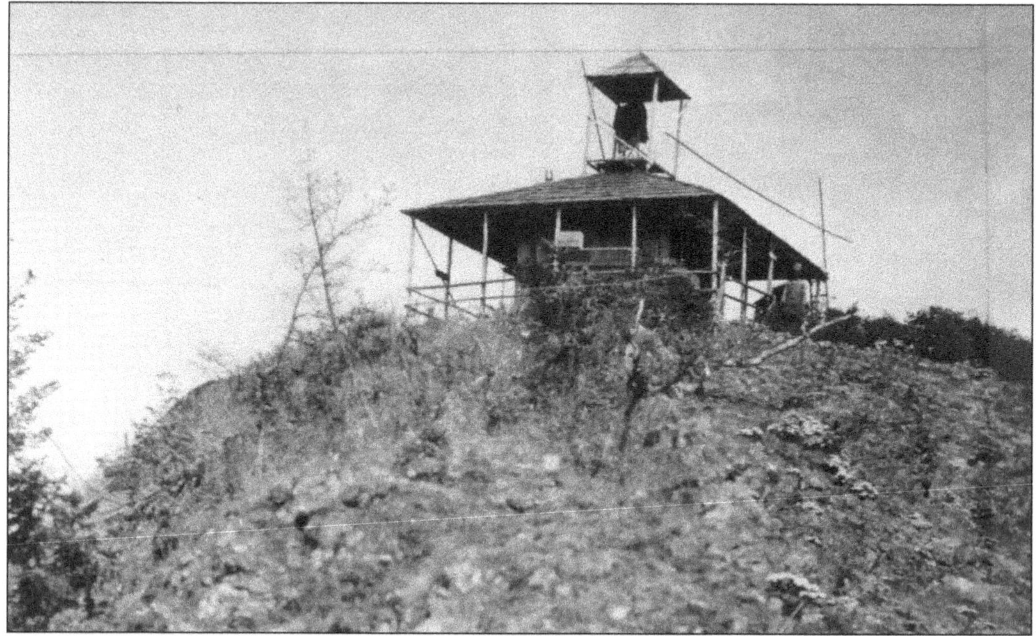

This unusual lookout was built on Heckletooth Mountain in 1917. The 10-by-10-foot cabin was covered by a 22-by-22-foot roof. An open-air cupola sat on top. The mountain got its name from the tall rocks near the summit that resemble the teeth of a heckle, an instrument used in handling flax.

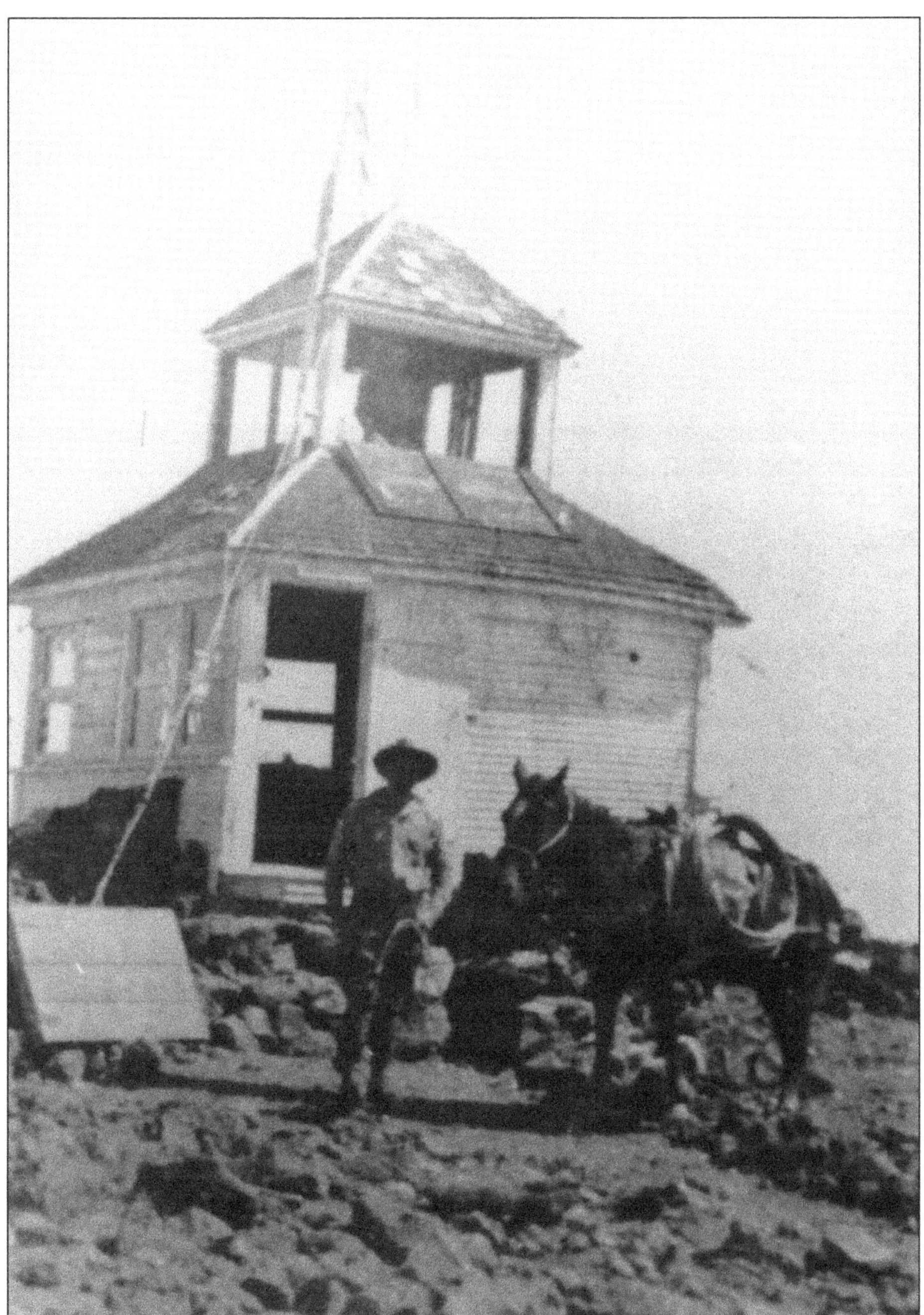

This 1920s photograph shows the first lookout that was constructed on 9,056-foot Bachelor Butte. Leila Hoover experienced an intense storm here in June 1925. The storm lasted three days and produced 60 mph winds. When it was all over, the cabin and mountaintop were encased in ice. Another time, she hiked down to a spot where a supply of fruits and vegetables had been cached for her. She left the cabin under a cloudless sky but was overtaken by a blizzard on her return trip, barely making it back to the cabin alive. The cupola cabin was replaced in 1931, but the new lookout was no longer staffed after 1939. The mountain became the site of a ski resort in 1958, and the name was changed to the loftier-sounding Mount Bachelor in 1983.

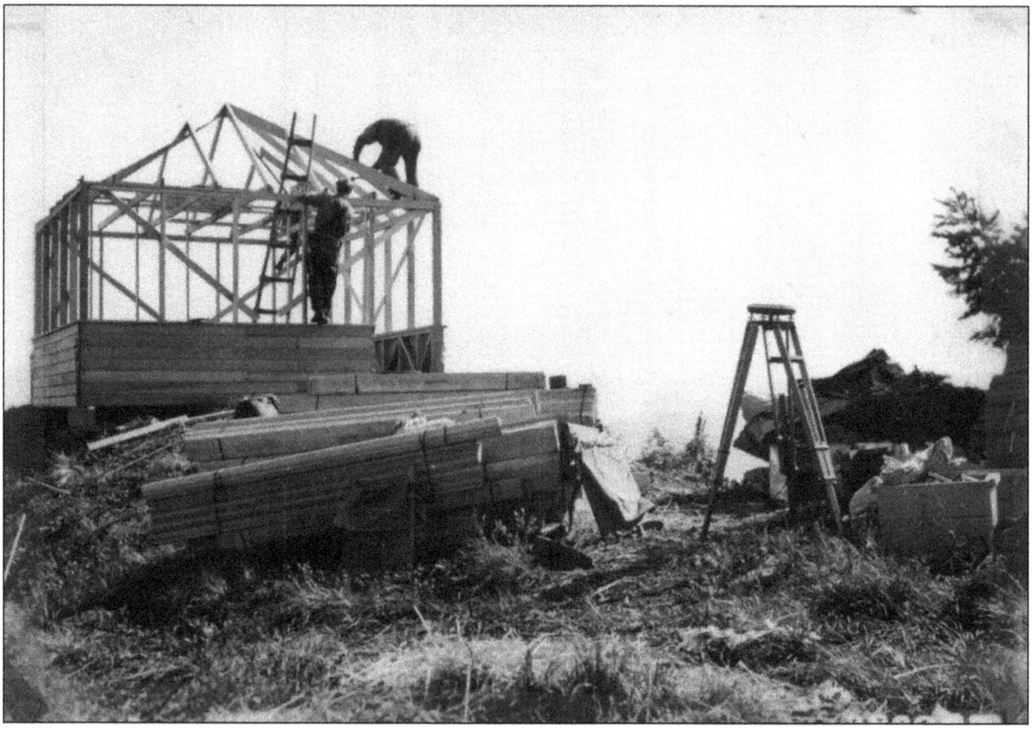

Two men work on the construction of the Henline Mountain lookout, which was finished in 1934. No road existed to the summit, so all materials and supplies had to be hauled in with pack animals. The lookout was removed in the 1960s.

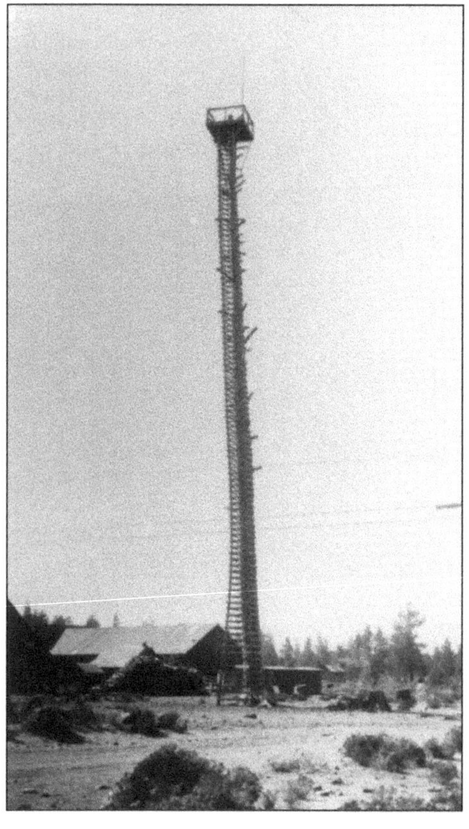

A 110-foot-tall ponderosa pine was home to a crow's nest lookout in the town of Sisters. A local blacksmith fashioned iron steps, which were driven into the trunk. When the tree died, it was cut down in 1926. (Forest History Society.)

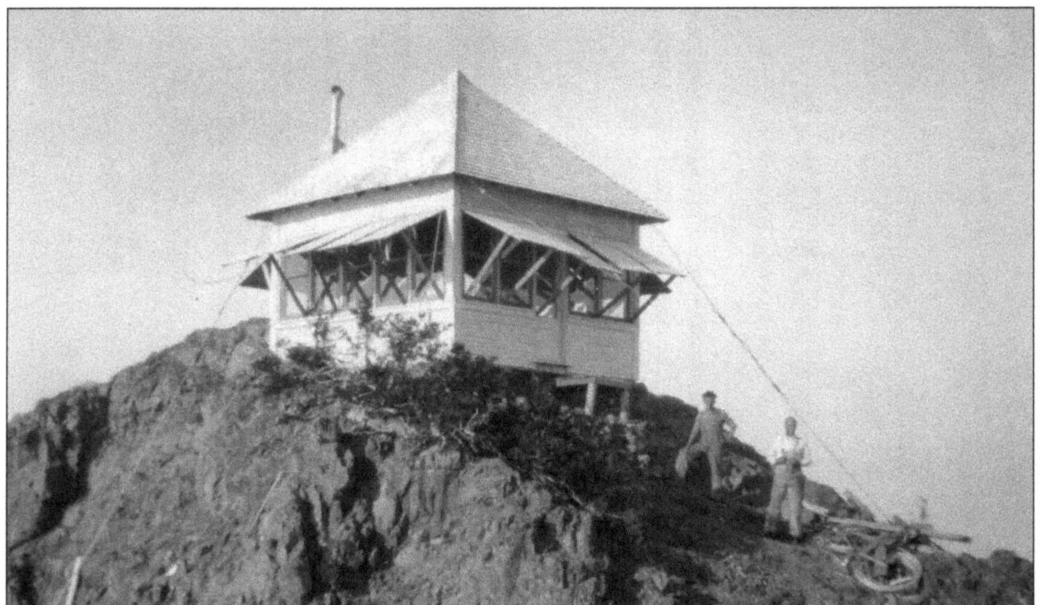

This lookout on Tidbits Mountain was constructed in 1926. Al Sorseth was stationed here in 1941 and remembers that, due to the lack of roads at the time, it was an 18-mile hike to get to the lookout. It was removed in 1967.

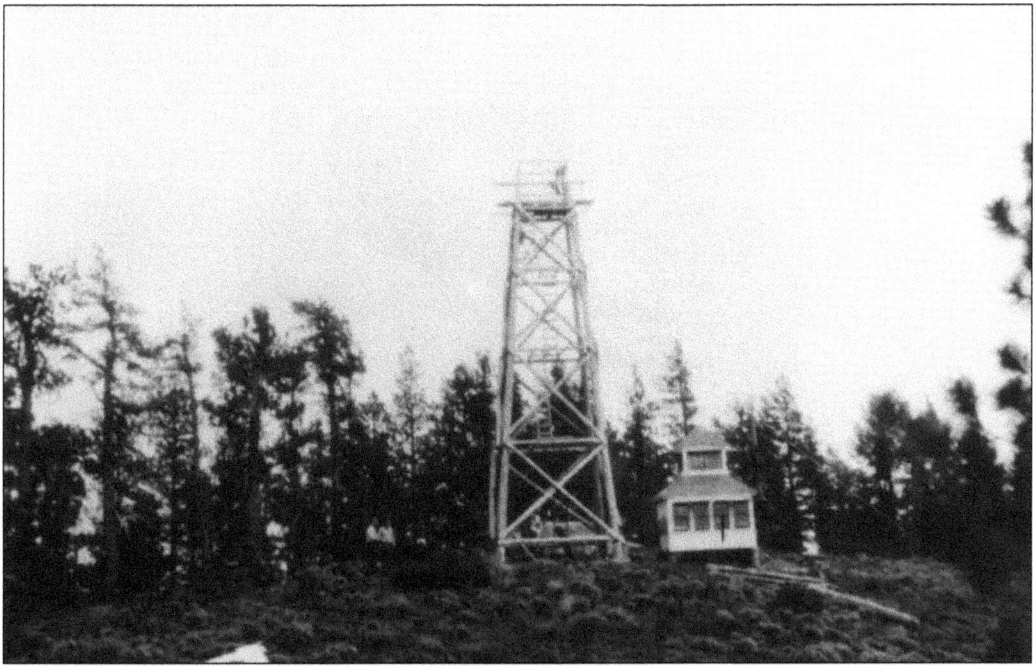

The 1920 cupola cabin and 1932 tower on Pine Mountain are shown in this undated photograph. Ernest Putnam was working at the Pine Mountain lookout in July 1927 when he decided to play his violin over the phone line for fireguard Clifford Martin. Unbeknownst to Putnam, his much-appreciated music was broadcast over the lines to all the ranger stations, lookouts, and fire camps in the Deschutes National Forest. Messages of congratulations came in from all over the forest after the performance.

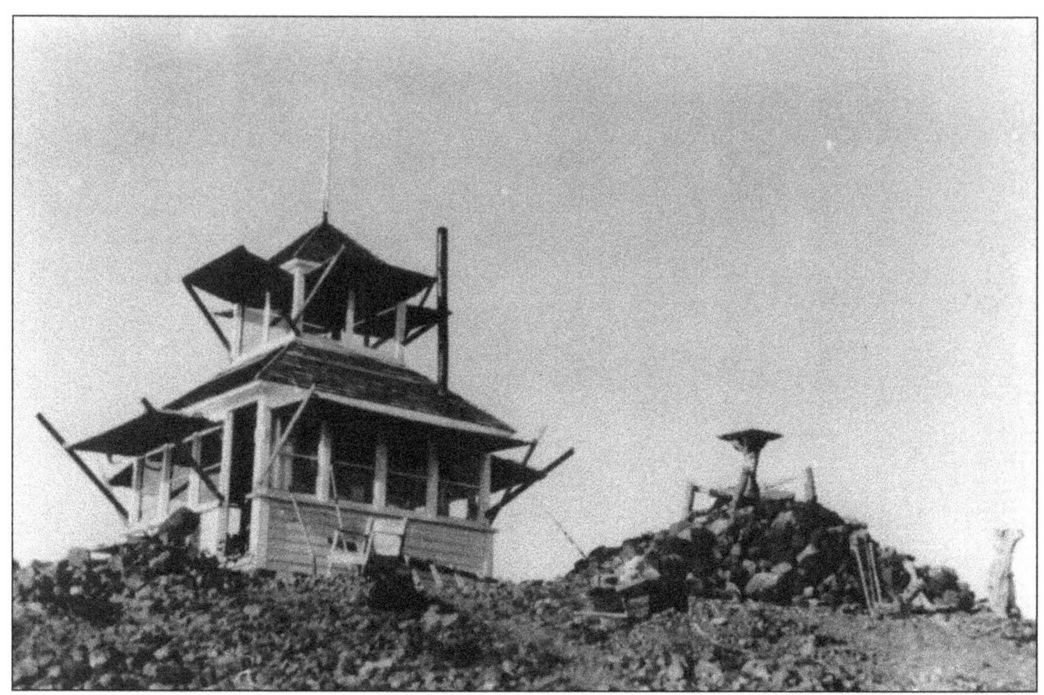

This cupola lookout cabin was built on Maiden Peak in 1923. Forest Service records indicated that the peak was the center of a "lightning fire belt," so in 1928, O.F. McMillan and E.L. Starr, experts in high-tension electricity, tested out a new device here. The device was designed to protect lookouts and telephone lines, both of which were susceptible to lightning strikes.

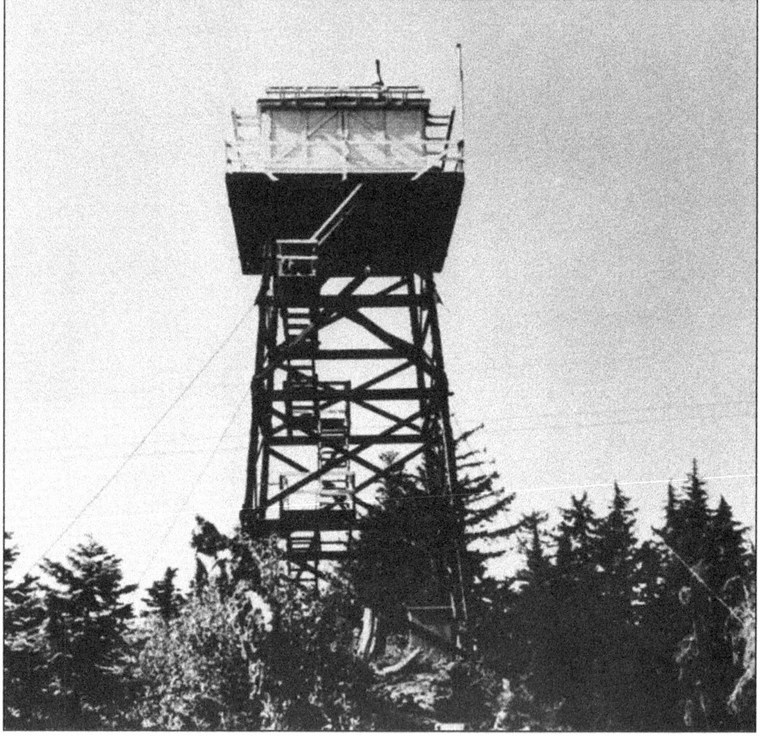

In 1953, a 52-foot-tall tower was constructed one mile southeast of the original 1930 ground cabin on Mule Mountain. Lindy Oleson was transferred to this lookout from Fuji Mountain in the summer of 1943, and he had to walk to get there. The Mule Mountain lookout leaked so badly that he ended up sleeping in the barn whenever it rained. (LCHM.)

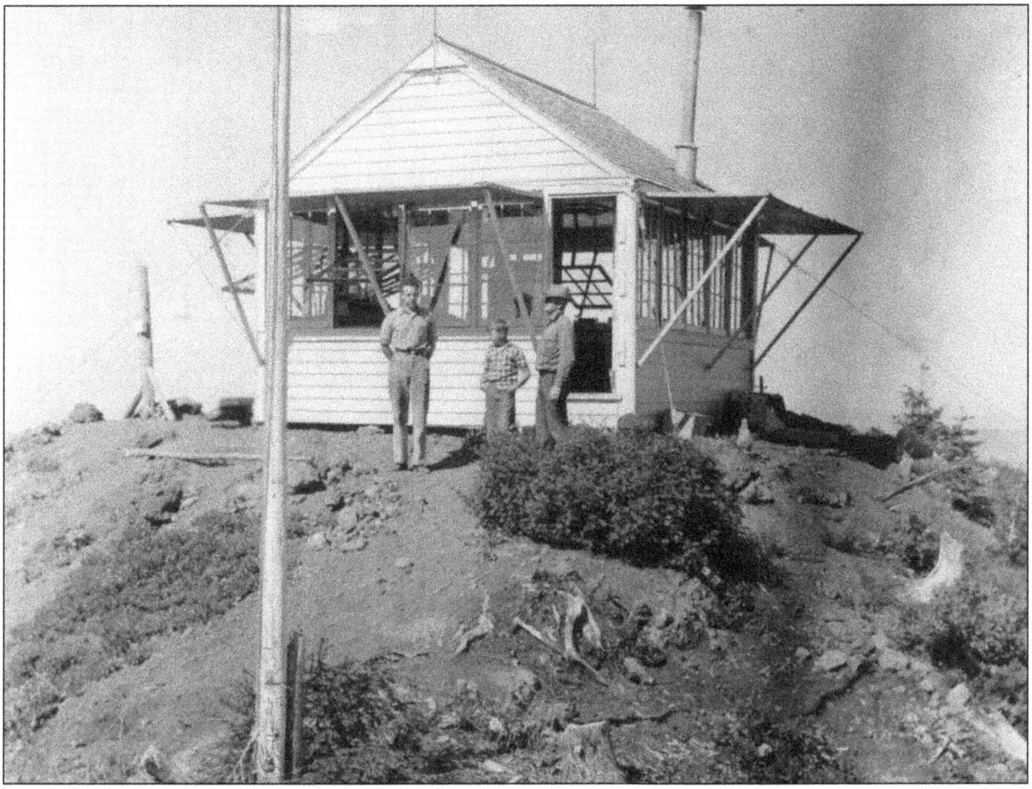

An L-4 lookout was built on Twin Butte in 1932. Al Sorseth had just moved to Sweet Home in 1938, escaping the Dust Bowl in North Dakota. He lucked into a job at Twin Butte that summer when the previous lookout was responding to a fire, fell off a cliff, and was too injured to return to duty. Sorseth's supervisor trained him how to do the job over the telephone.

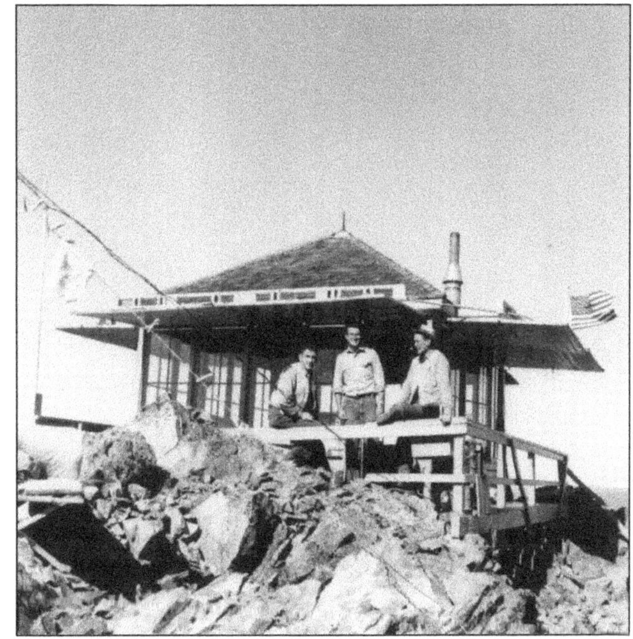

One summer, a college student stationed at Dome Rock was eager to get down at the end of the season. It looked likely to rain, meaning he could go home. So, he chucked all of his remaining groceries over a cliff. When the rain failed to materialize, he had to stay a few more hungry days.

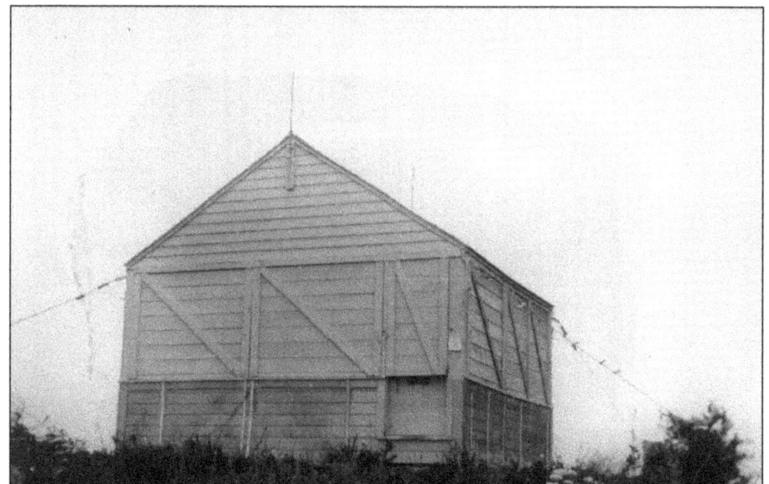

An L-4 lookout cabin was built on Olallie Mountain in 1932. It was last used in the 1970s and has stood abandoned since then. Along with Rebel Rock, it is one of only two lookouts left in the Three Sisters Wilderness.

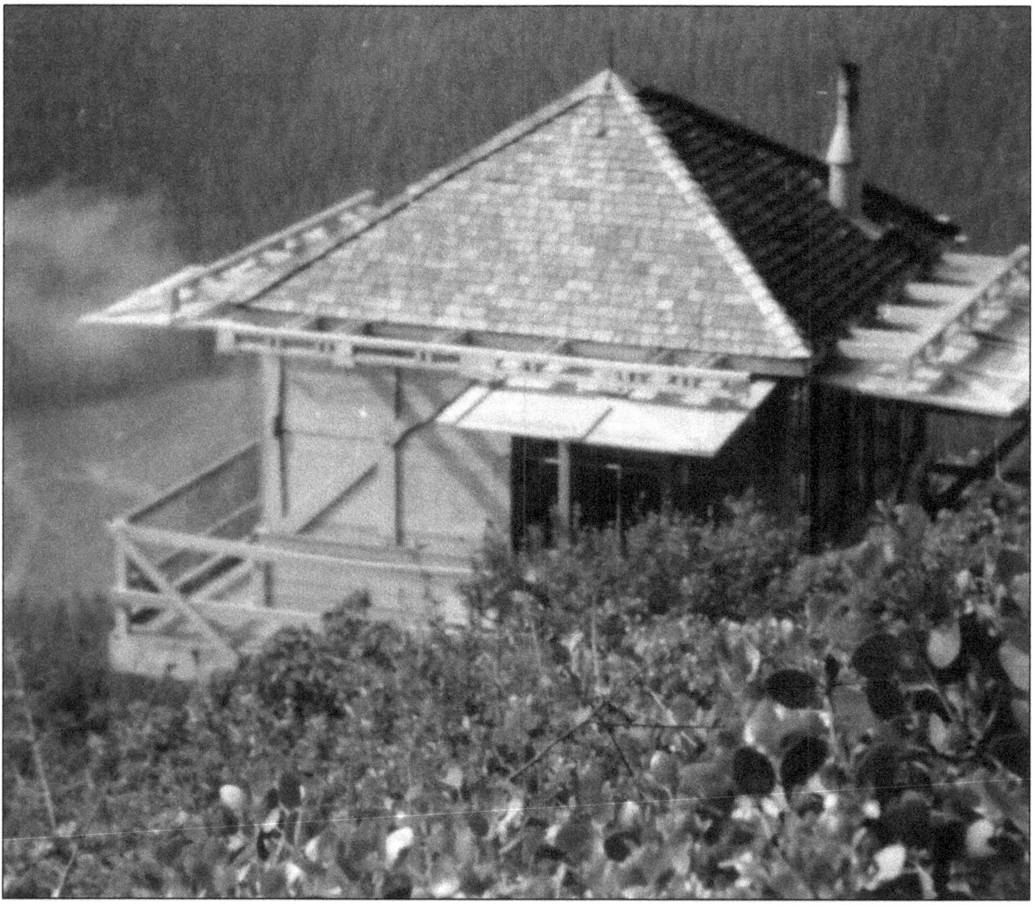

Jim Drury recalled one summer during his career with the Willamette National Forest when the lookout man at Rebel Rock wanted to quit and come down. Drury met him at the trailhead. "Here, he came down the road, and he was beating Prefontaine all to hell when it came to speed." Drury gave him a ride, but as soon as they reached the highway, the young man jumped out and ran off.

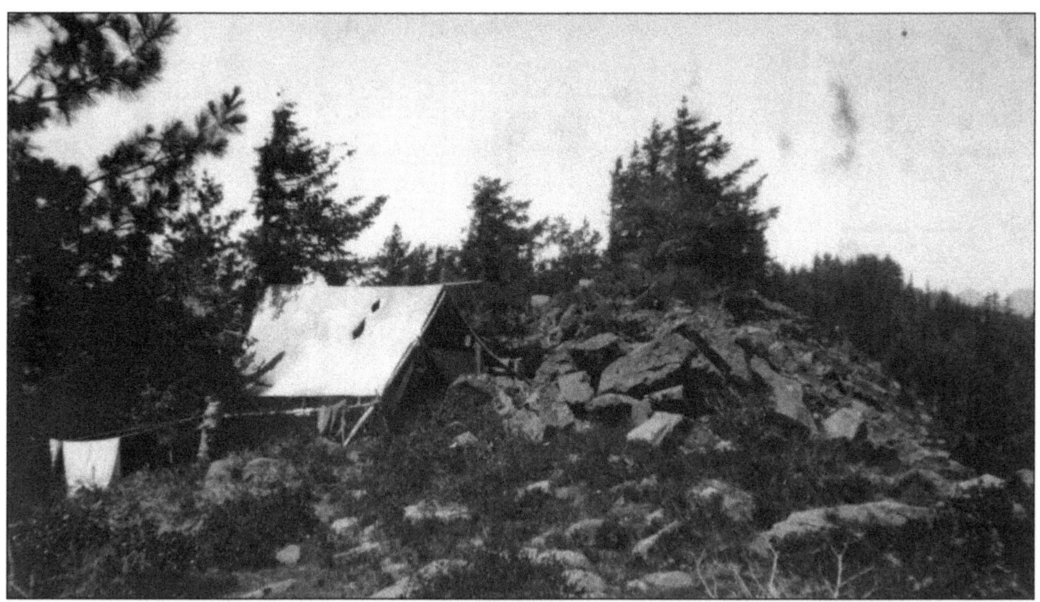

A lookout camp was established on Horsepasture Mountain in 1915. University of Oregon student Martha Andrews was stationed at the camp starting in 1919. Her sister Dorothy was stationed at nearby Frissell Point, which was also a tent camp. They could see each other with binoculars, and when they got lonely, they could talk to each other via the phone system. Pictured above in 1924 is a tent camp on Frissell Point. Below is the D-6 cupola that was built on Horsepasture Mountain in 1922. The same type of cabin was built on Frissell in 1926. Both lookouts were removed in the 1960s.

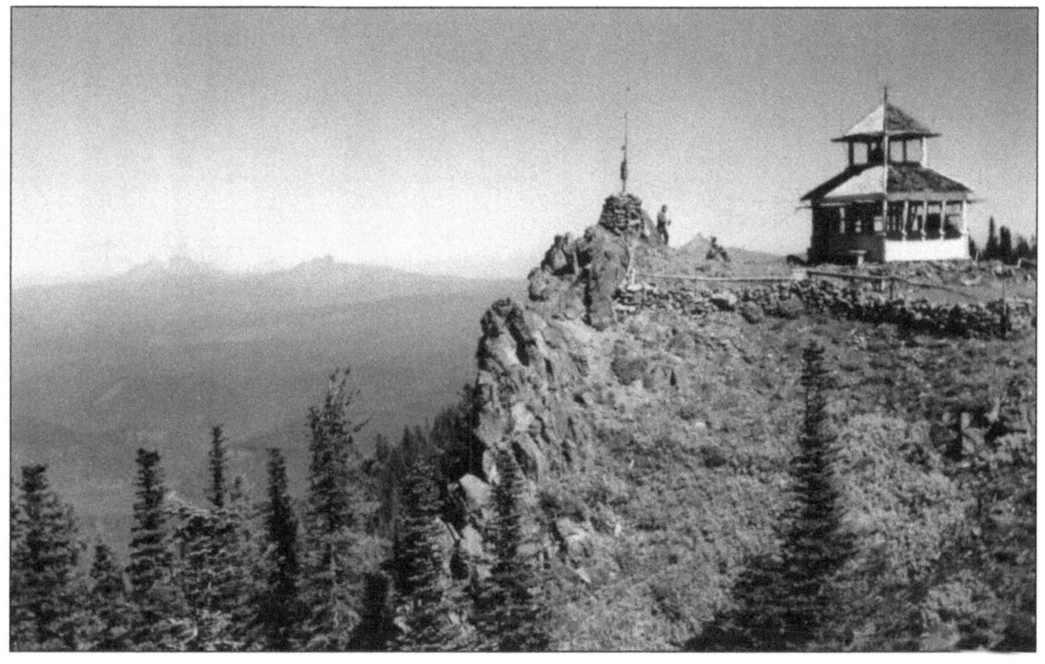

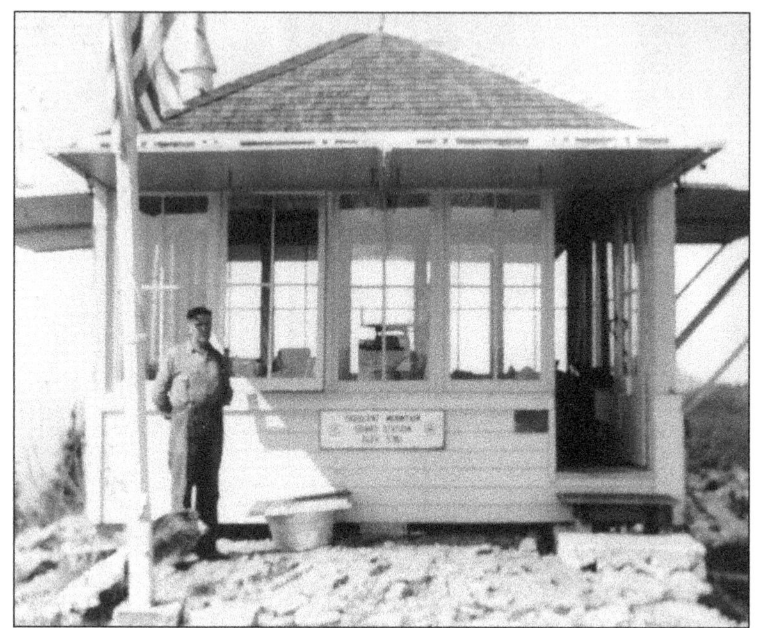

An L-4 lookout cabin (pictured) was built on Crescent Mountain in 1938, following a 1922 cupola cabin. In 1945, 16-year-old Douglas Lindsay (not pictured) was stationed here. He was found dead from a gunshot wound on the front steps of his cabin on July 30. An investigation determined it was self-inflected but from accident or suicide, no one knew.

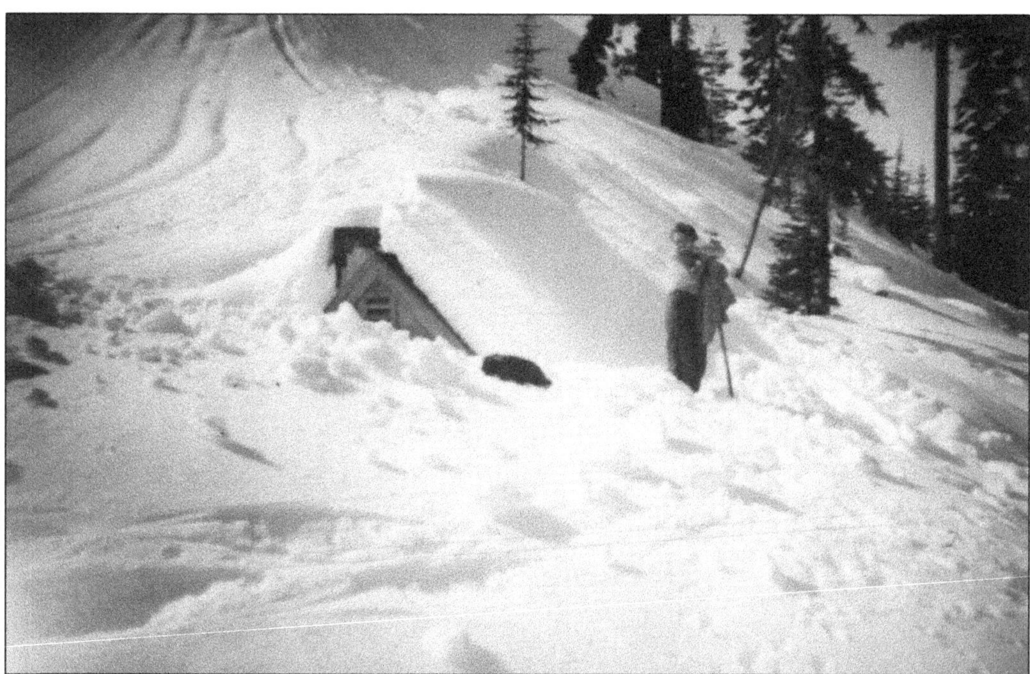

Manena Schwering and Mrs. Buck Merriot were hired by the Aircraft Warning Service to spend the winter of 1942–1943 on Gold Hill. They were on alert for enemy aircraft around the clock and arranged it so they alternated six hours on and six hours off. They got very tired of their food and came across a stash of unlabeled cans left by the crew that had built the cabin. They were excited for something new in their diet, but when they opened the cans, they discovered roofing tar.

The Minniece Point lookout (left) and Ranger Glenn Charlton (right) stand outside the lookout cabin in this 1939 photograph. The L-4 lookout was built in the Willamette National Forest in 1934. The mountain was named after John Minniece, an early Forest Service employee for the Detroit Ranger District.

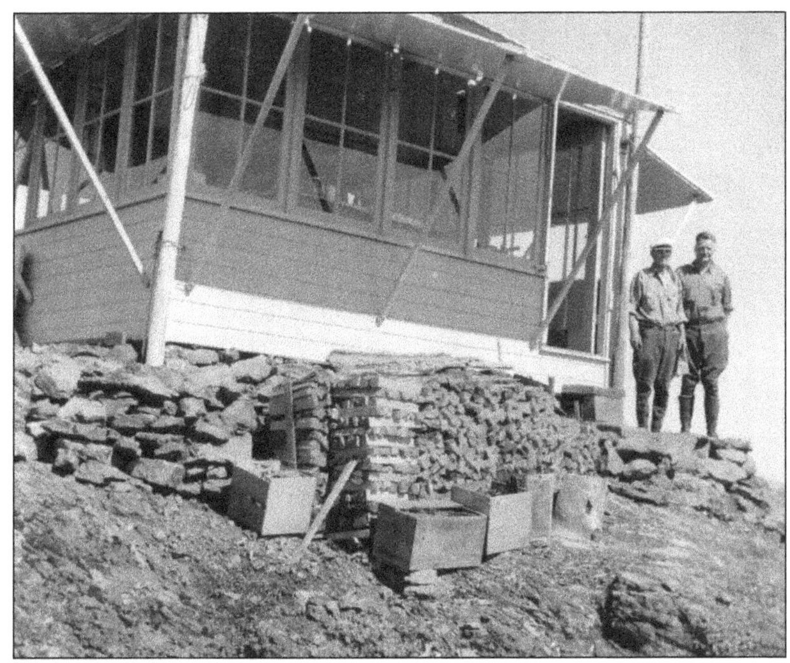

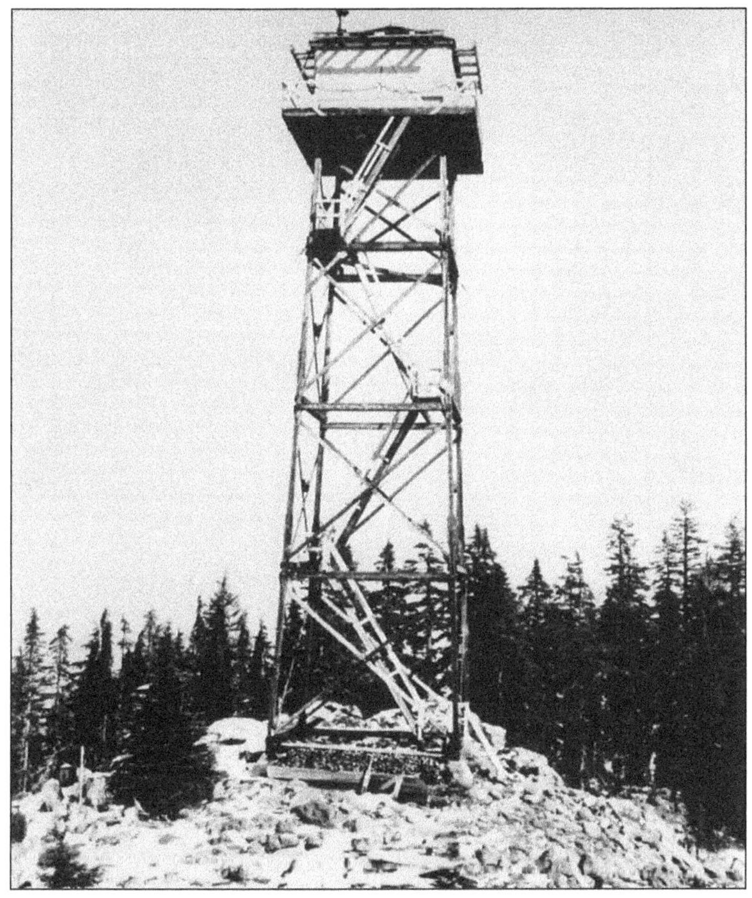

A 60-foot-tall tower was built on Wolf Mountain in 1931, replaced by a 65-foot-tall tower in 1940. When Harold McKenzie was stationed on Wolf Mountain, he made friends with a chipmunk that became his pet. He made a little copper collar for it and a leash with a string. When the chipmunk died, he buried it with a gravestone. (LCHM.)

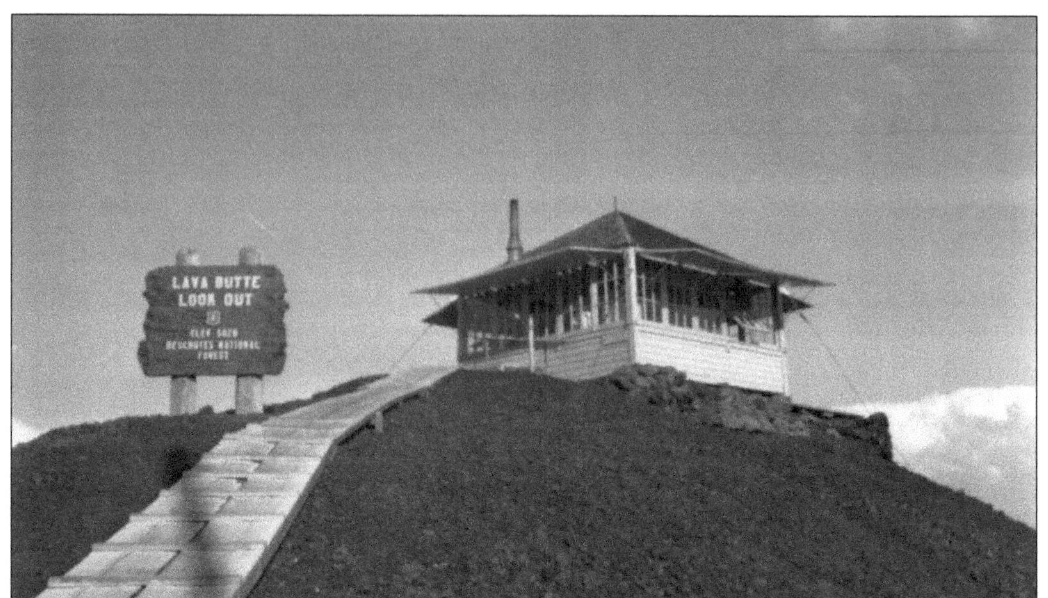

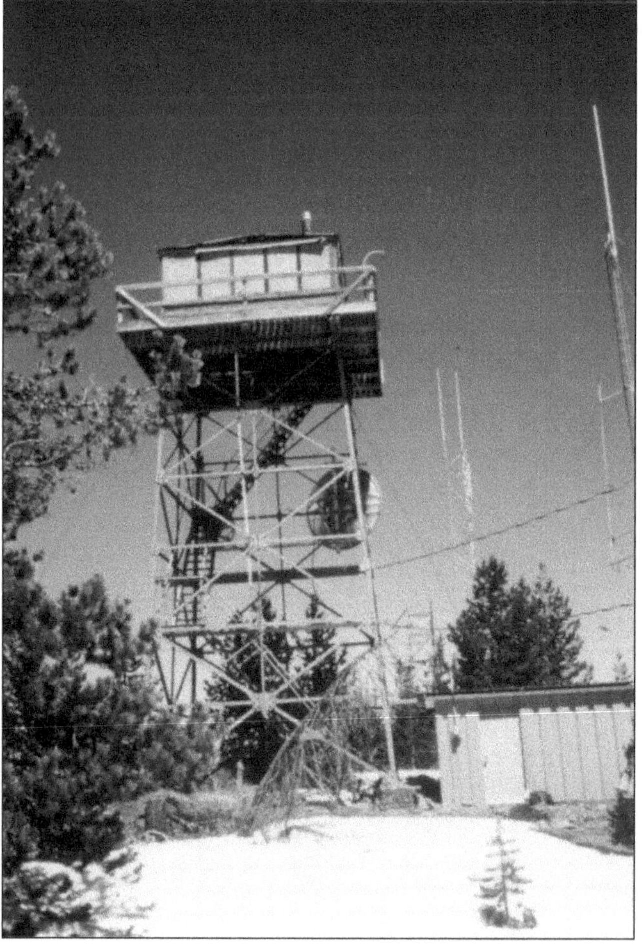

The cinder cone known as Lava Butte is capped by a crater that is the result of a single eruption about 7,000 years ago. This L-4 lookout was built on the highest edge of the crater rim in 1931. The current lookout, built in 1998, is still staffed and has a small ground-floor visitor center. Lava Butte now sits within the boundaries of the Newberry National Volcanic Monument, which was created in 1990.

Walker Mountain was one of four lookout sites established in the Deschutes National Forest in 1913. The original lookout was a crow's nest, and a stone cabin was built as living quarters around 1917. A 35-foot-tall steel lookout tower was built in 1932. The lookout is only used as a communication tower now, but the cabin was restored in the 1990s.

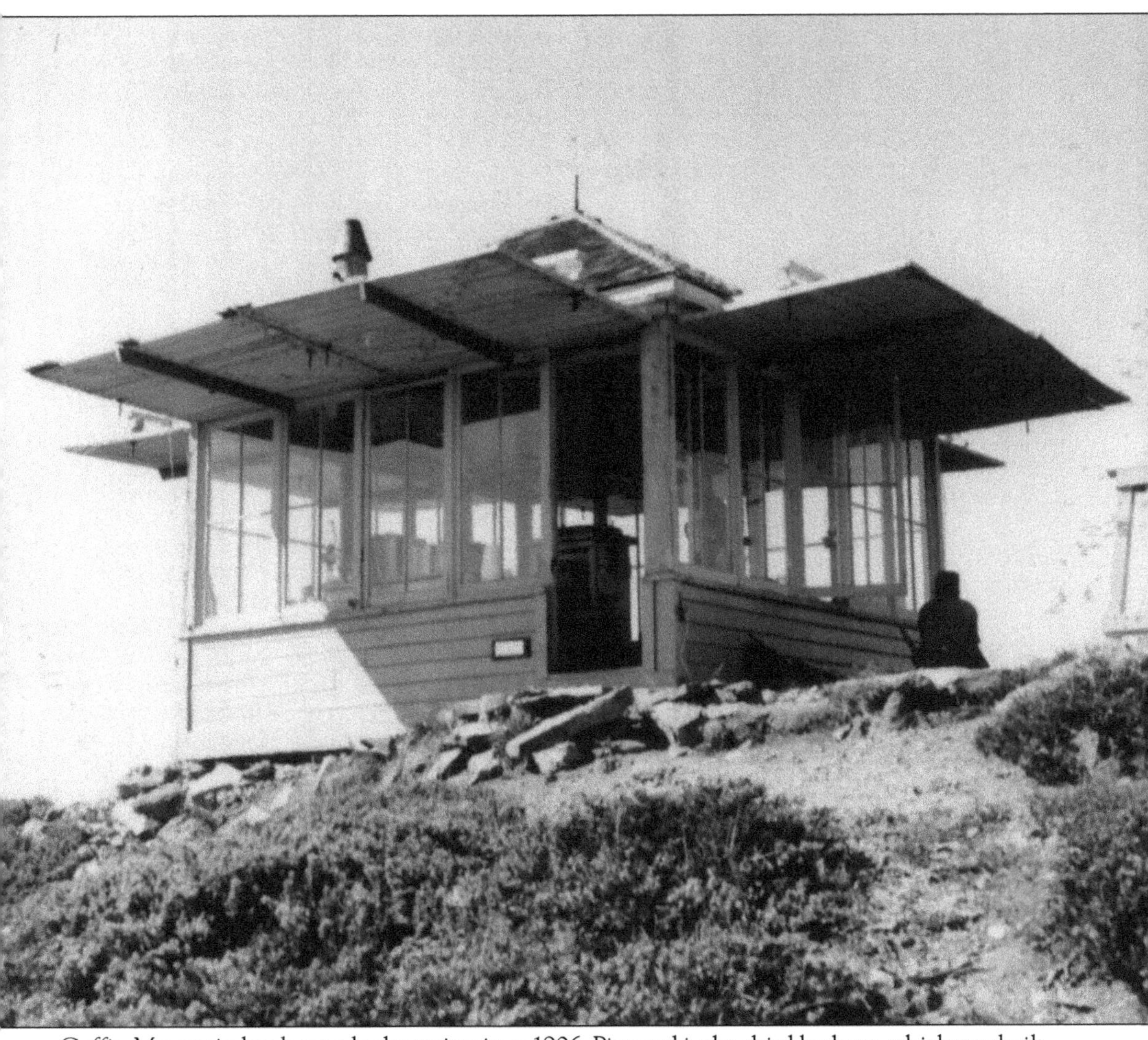

Coffin Mountain has been a lookout site since 1906. Pictured is the third lookout, which was built in 1936. In 1986, it was replaced by the D-6 lookout that is still staffed today. Ann Amundsen has worked on Coffin Mountain since 1990, and despite the steep hike to get to the summit, she estimates she sees 500 hikers a summer.

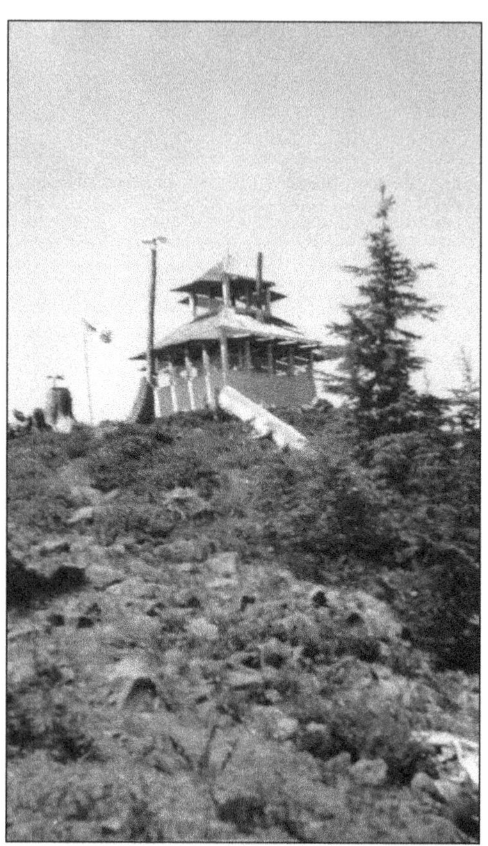

In 1944, Charles and Melvina Neet were stationed at Waldo Mountain lookout for only three weeks before getting a new job offer. District Ranger C.B. McFarland needed a packer and offered the job to Charles, who at first declined. When he finally agreed, it was on the condition that his wife could come with him. He later reported "We never had a better summer in our lives. It was wonderful!" Pictured is the second lookout, built in 1929.

Hurricane-force winds toppled thousands of trees in the valley of the Metolius River on April 22, 1931. Plans were made to salvage the 10 million board feet of timber, but the Forest Service was fearful that the downed trees would create a fire hazard before they could be removed. So a lookout was established on Cache Mountain to help watch for fires in the area. The lookout and garage are pictured in 1958.

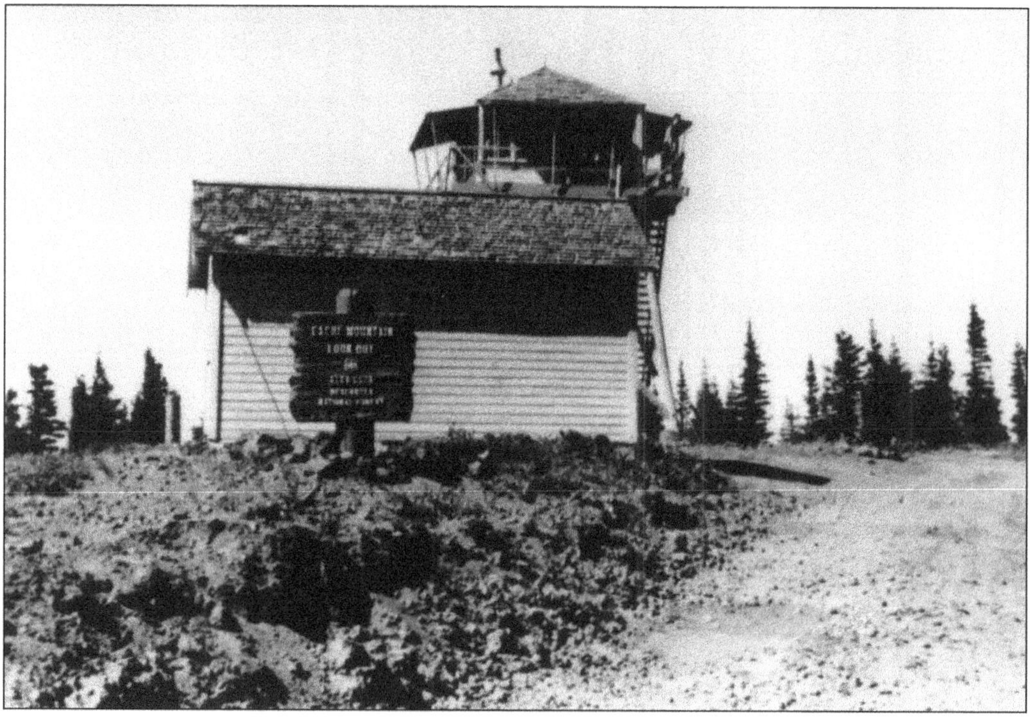

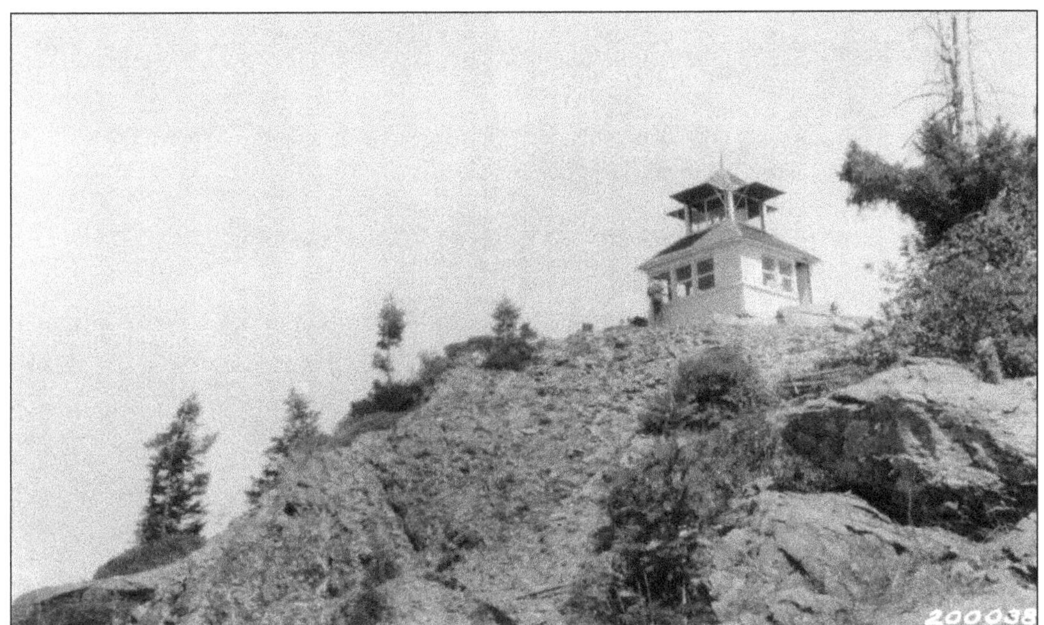

This cupola lookout on Castle Rock was built in 1926 and was later replaced with an L-4 cabin in 1938. Access was by a steep 4.5-mile trail, but in the 1960s, the Forest Service built a road so close to the mountain that the hike was shortened to a mile. By that time, the lookout was no longer staffed, and with the easier access came inevitable vandalism. Vandals burned the cabin down on October 31, 1974.

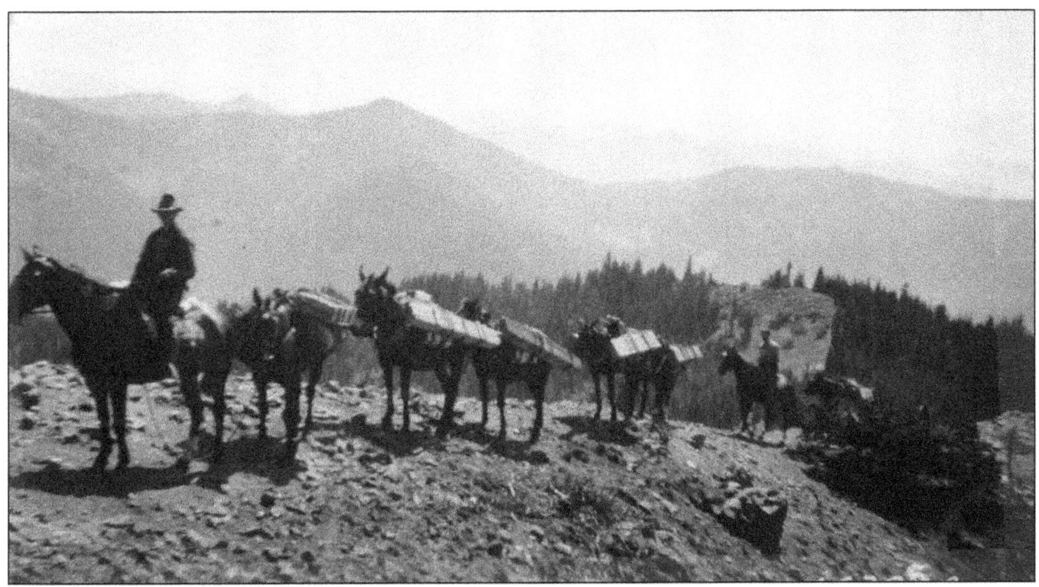

Many lookout sites were not accessible by road and were only reachable via narrow winding trails. In such cases, pack animals were used to get supplies to the summit, as shown in this 1922 photograph of a pack string on the way to Castle Rock.

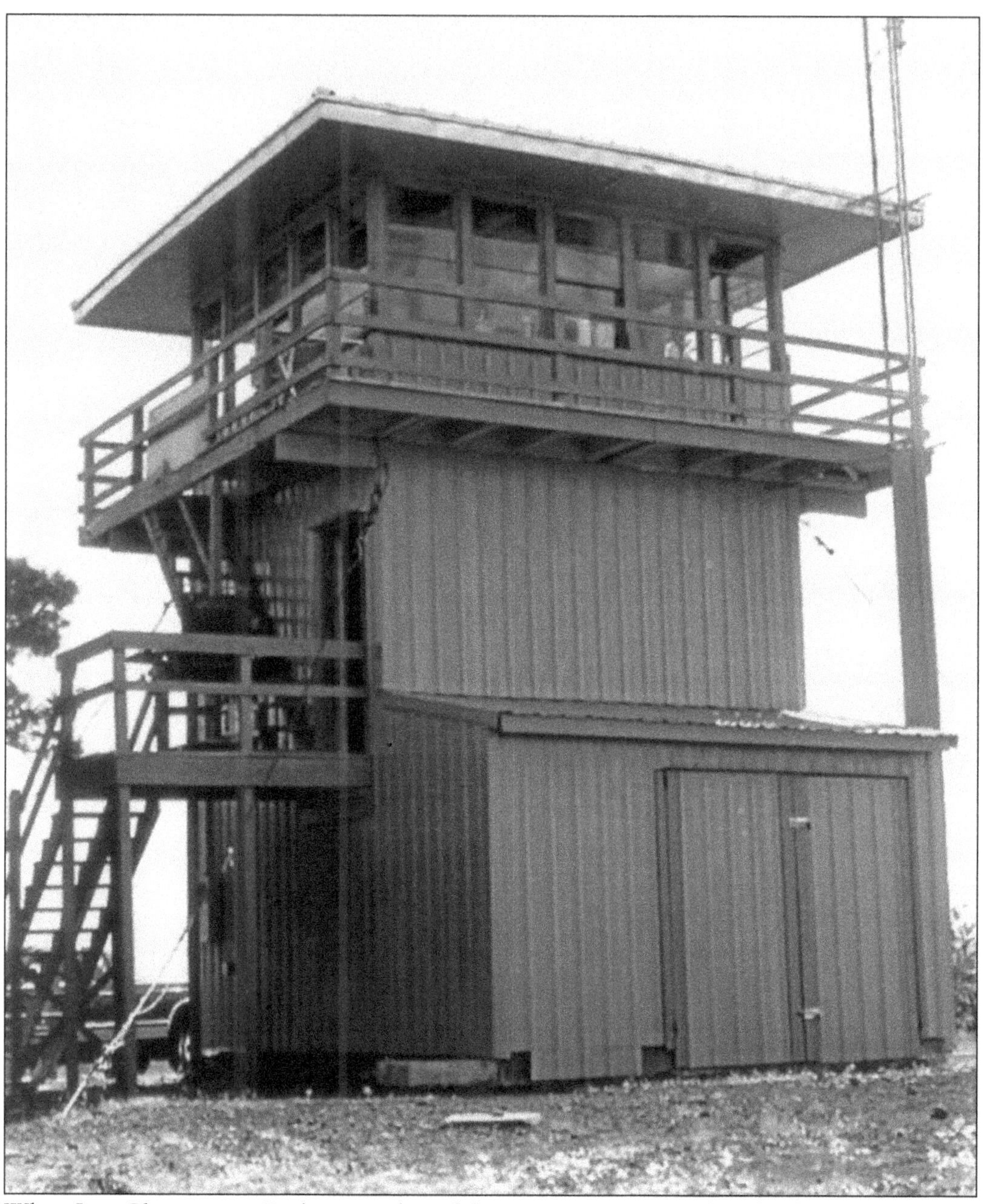

When Lois Christiansen Eagleton graduated from Junction City High School in 1952, she was looking for a summer job to help pay for college. She landed a job as a lookout on Henkle Butte, near the town of Sisters. During her first sleepless evening in the tower, the wind blew all night, violently shaking the lookout. When she reported this the next day, a crew was sent out to check on things. They discovered that many of the nuts and bolts that held the structure together were loose. After they were tightened, the wind never bothered Eagleton again. In 1961, the 40-foot-tall wooden tower that Eagleton lived in was replaced with the lookout pictured. Even though the butte is named for volunteer infantryman Jeremiah E. Henkle, it is sometimes referred to as Hinkle Butte. (Ron Kemnow.)

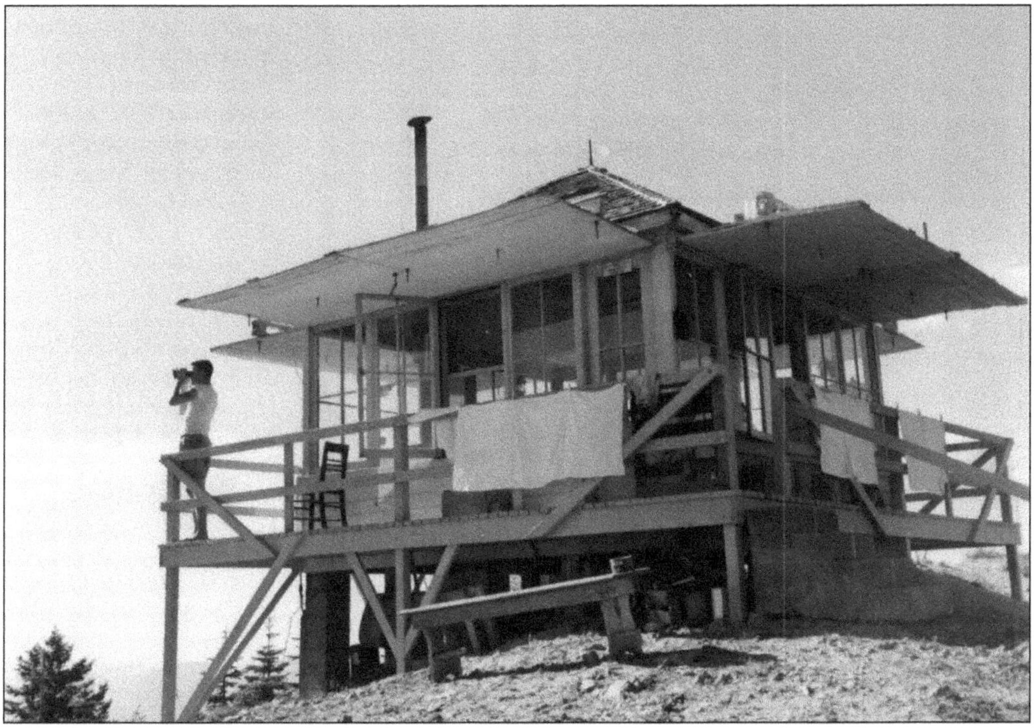

This 1962 photograph shows Dave Shetzline on the catwalk of the Huckleberry Mountain lookout in the Willamette National Forest. The L-4 lookout was built in 1939, and by the 1990s, it was in bad shape. Volunteers spent three years restoring it, but right when they finished in 1994, vandals kicked in the front door, set a few small fires, stole the fire finder, and left behind drug paraphernalia. Repairs were made, and the lookout is staffed every summer.

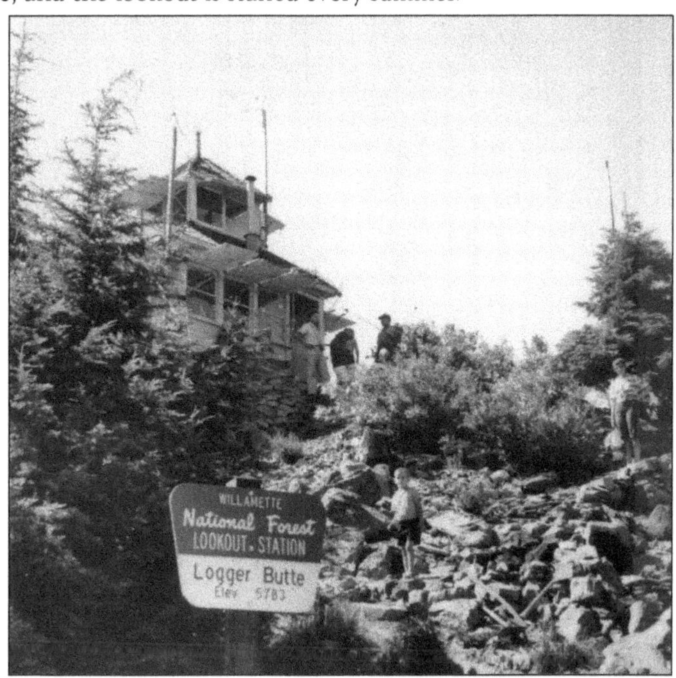

A D-6 cupola lookout was built on Logger Butte in 1926. Two years later, when E.E. Kilpatrick and his wife, Iva, were stationed here, she hiked down the trail to get water and encountered a cougar. She dropped to one knee and shot the cougar through the heart with her .22 caliber rifle before continuing on to fetch the water. The cougar was later measured and found to be eight feet long.

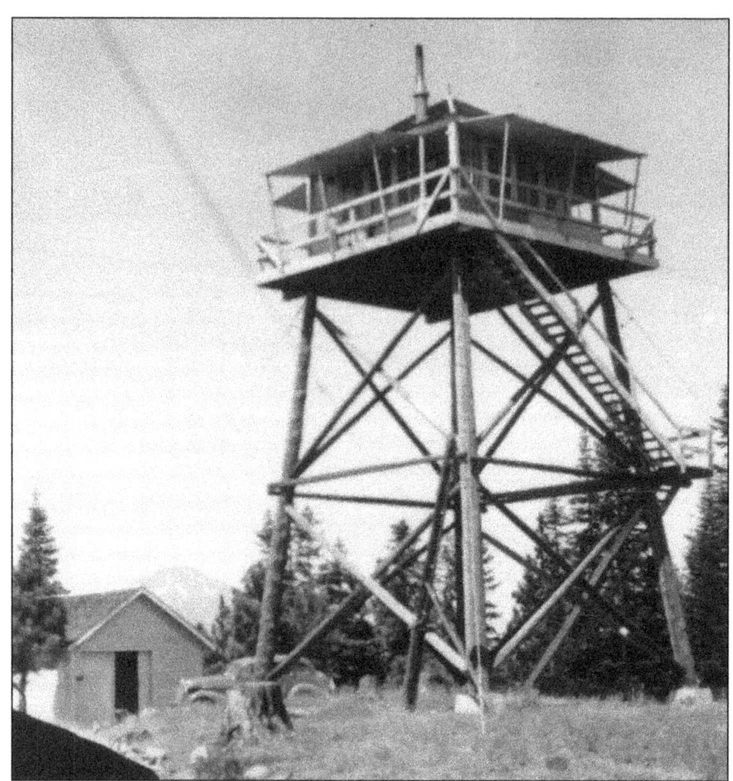

Lee Smith was stationed in a tent on Wanoga Butte during the summer of 1932 while a lookout tower was being constructed. A mother bear and her two cubs raided his tent twice in less than 48 hours, destroying all of his food supplies and scattering a kettle of cooked beans all over the summit. It was reported in the *Medford Mail Tribune* that Smith eagerly awaited the completion of the bear-proof tower.

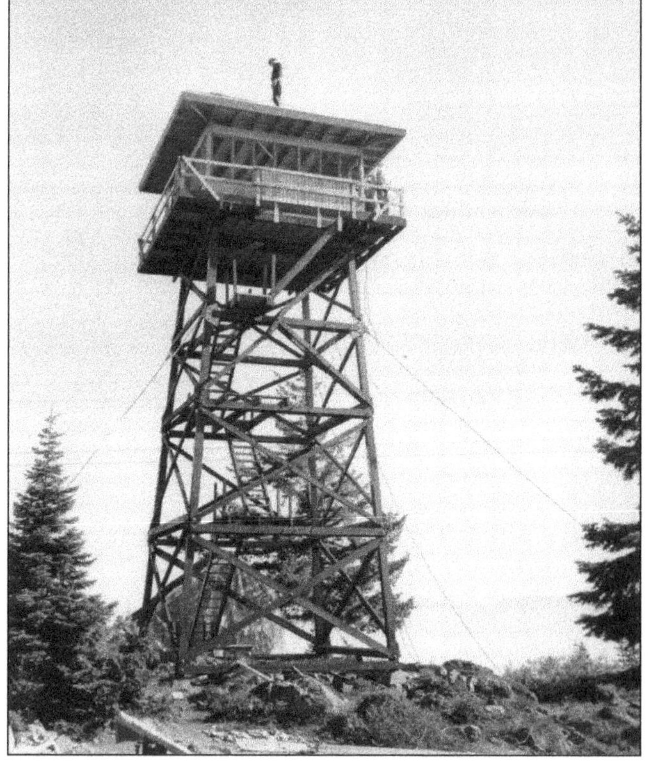

Grass Mountain became a tent camp lookout site in the 1920s, followed by a 40-foot-tall tower in 1933. Pictured is the last lookout in the final stages of construction in 1962. The lookout was removed in 1982, but the salvaged material was used to build a new lookout on Warner Mountain in 1986.

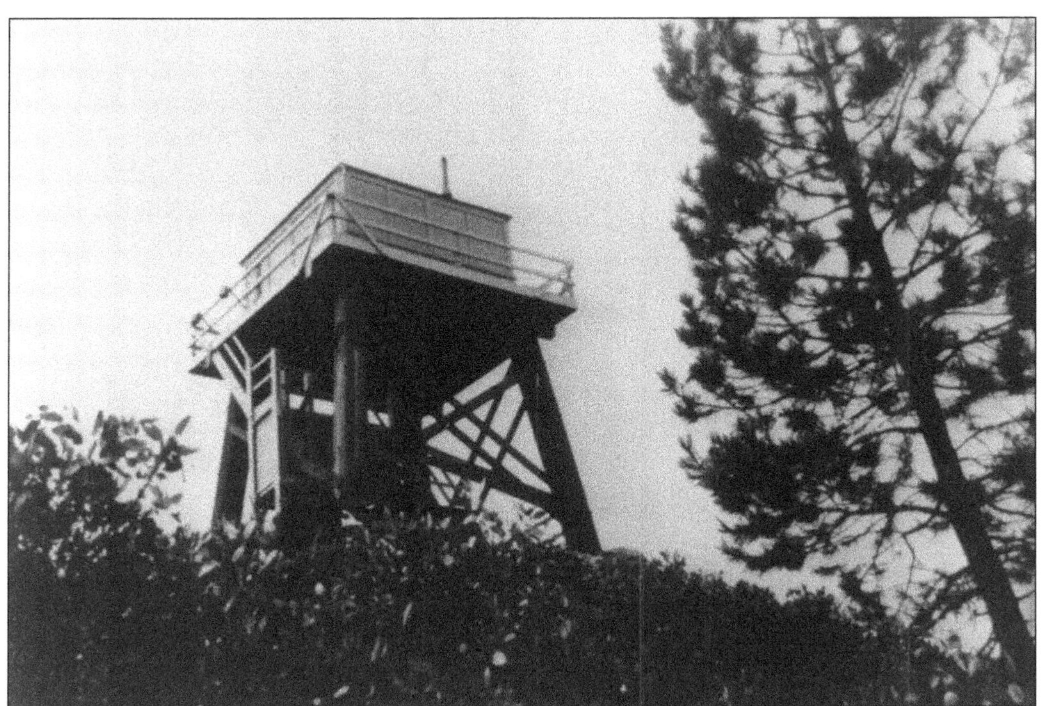

The lookout on Pistol Butte was not yet complete in late September 1932 when a wildfire near Wickiup Butte started. Without a fire finder on the butte, the fire could not be accurately charted, so the Forest Service put out a call for assistance to the Walker Range Protective Association (WRPA). When two WRPA men approached the fire, they found it was on the other side of the Deschutes River. The *Bend Bulletin* reported that "one of the pair arrived at the fire garbed as a nudist. The other plunged into the river fully dressed."

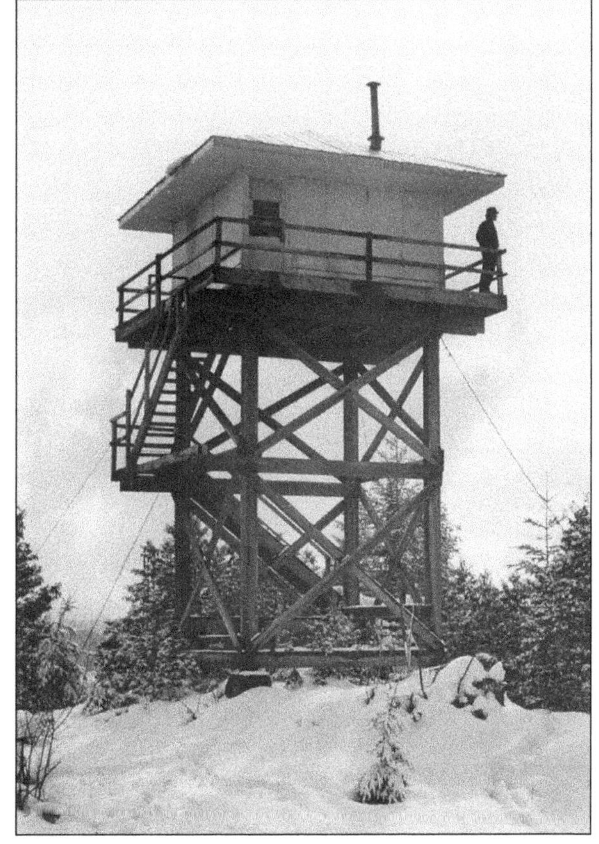

This 25-foot-tall L-4 tower was built on Mount Hagan in 1955. Vicki Timm was stationed here during the summers of 1971 and 1972, and the pay was $1.65 per hour. She brought a box of books with her and finished them all within the first two weeks. She also brought her dog and a big tomcat. The cat disappeared once but turned up again 10 days later, a little hungry but otherwise unscathed. The lookout was removed in 1979. (LCHM.)

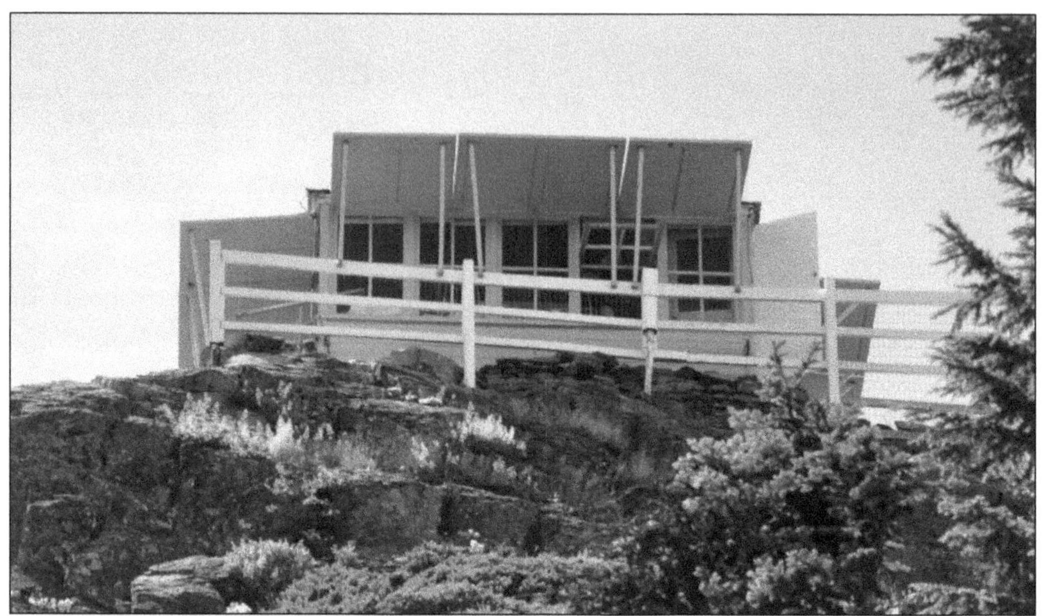

An L-4 lookout cabin was built on Carpenter Mountain in 1935, replacing an earlier lookout that had been destroyed by lightning. After falling into disrepair, it was restored in the 1990s and is now staffed every summer. It sits within the boundaries of the H.J. Andrews Experimental Forest. (Author's collection.)

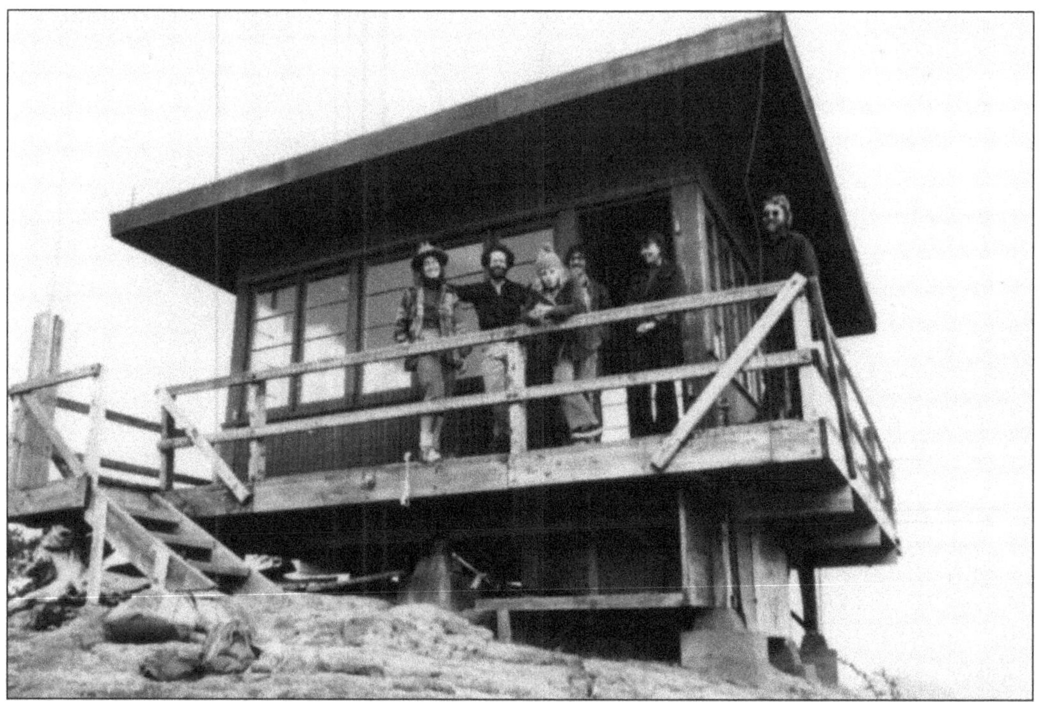

A D-6 cupola cabin was built on Little Cowhorn Mountain in 1923, followed by an R-6 lookout in 1960, pictured in 1987. Walter Wentz remembered riding out a thunderstorm when he was stationed here. The wind was intense, rain blew in through the cracks around the windows, and lightning struck about 200 yards away. The lookout is no longer staffed but was restored in 2004.

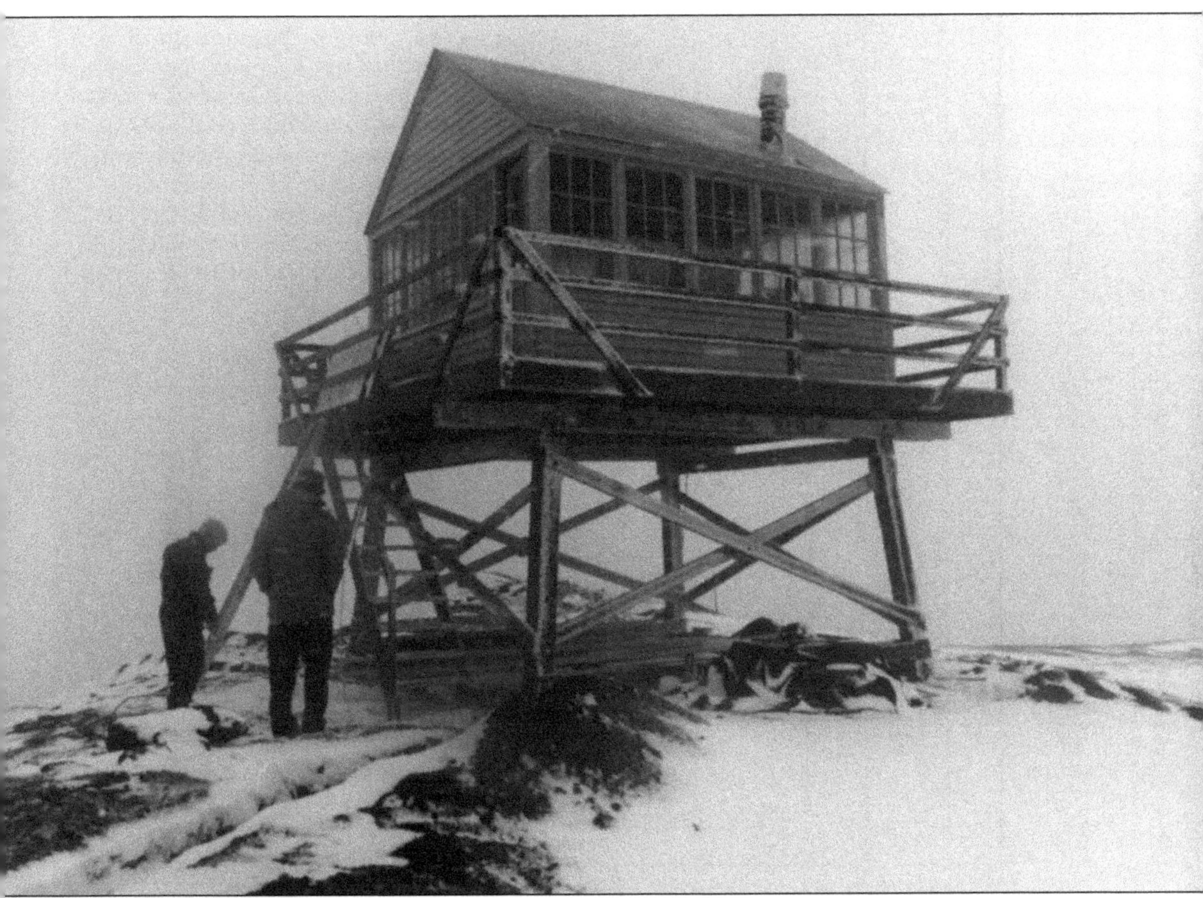

Sand Mountain has been a lookout site since the 1920s. After the lookout accidentally burned down in 1968, the site was used only intermittently. In the late 1980s, a group of volunteers worked with the Forest Service to reestablish a lookout on Sand Mountain, and in 1989, the L-4 lookout from Whiskey Peak in southern Oregon was salvaged for reuse. This photograph shows two men doing some last-minute work that fall, with winter snow already on the ground. The project spawned the creation of the Sand Mountain Society, a volunteer-based nonprofit that helps salvage and restore lookouts. Since their creation, they have participated in 11 full-blown restoration projects.

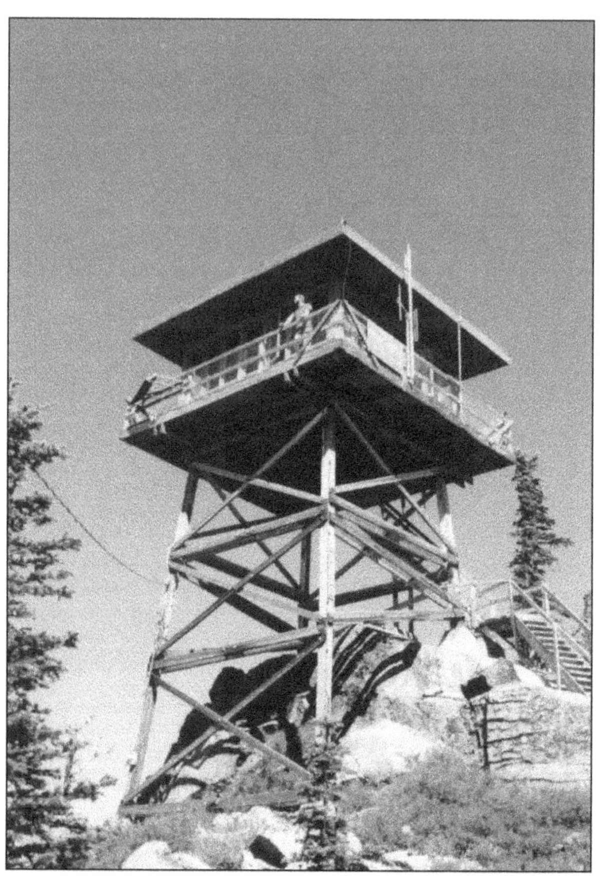

Odell Butte has been the site of three different lookouts since 1916. While staffing this lookout in 1930, Jack Benson found himself with a major rat problem. He tried several different kinds of traps without success until he devised a trap out of a cut-up tin can, which did the trick, allowing Benson to have restful nights once again. This undated photograph shows the current tower, which was built in 1963 and is still staffed every summer.

An L-4 lookout was built on Iron Mountain by the Civilian Conservation Corps in 1933. They used 200 sticks of dynamite to blast a level site for the building. The lookout was destroyed by wind and snow in 1976, so the unused lookout from Herman Peak in the Siuslaw National Forest was moved here. Pictured in 2007, the rundown cabin was removed that same year. (Author's collection.)

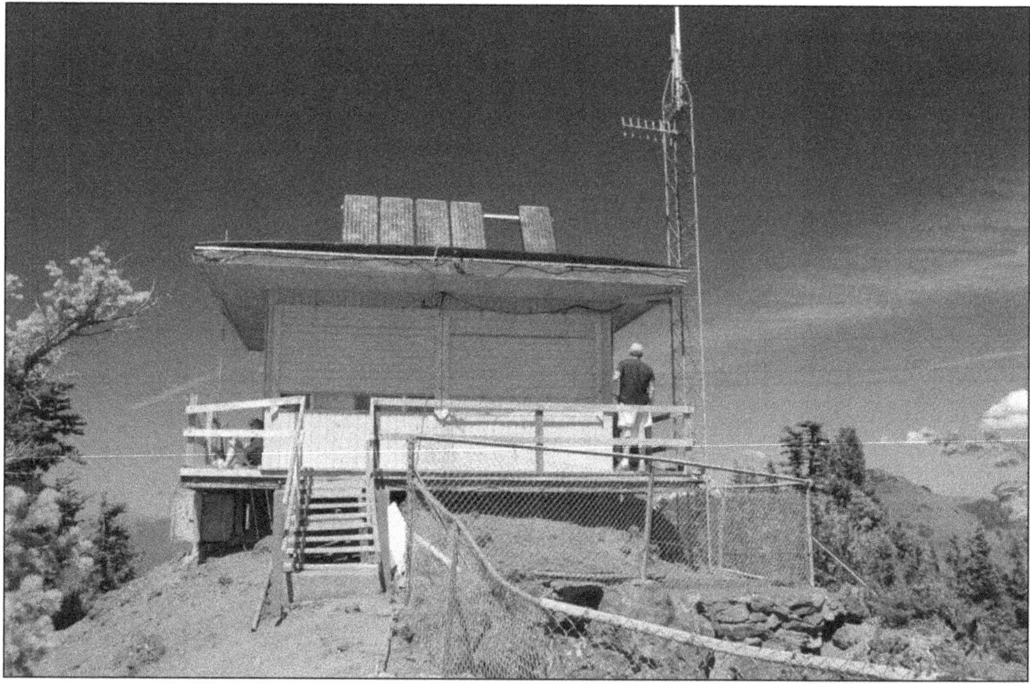

Spring Butte was first used as a lookout in 1931, with only a tent and a fire finder. C.E. Hein spent the summer on the butte and, with the aid of a 16-year-old assistant, was directed to clear all the timber that blocked his view. The assistant proved to be more hindrance than help with the saw, and Hein later reported in a 1971 *Bend Bulletin* article that the boy's "prowess as a cook was no better than on the saw." This 1991 photograph shows the 30-foot-tall lookout that Hein helped construct in 1932. It has since been replaced by a new tower.

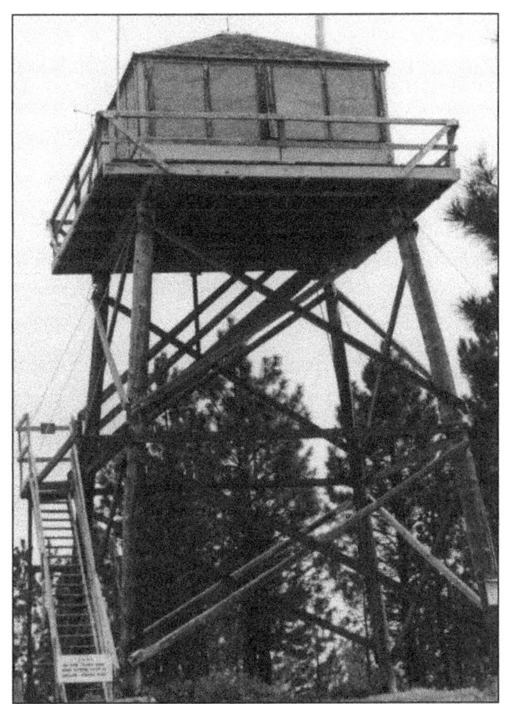

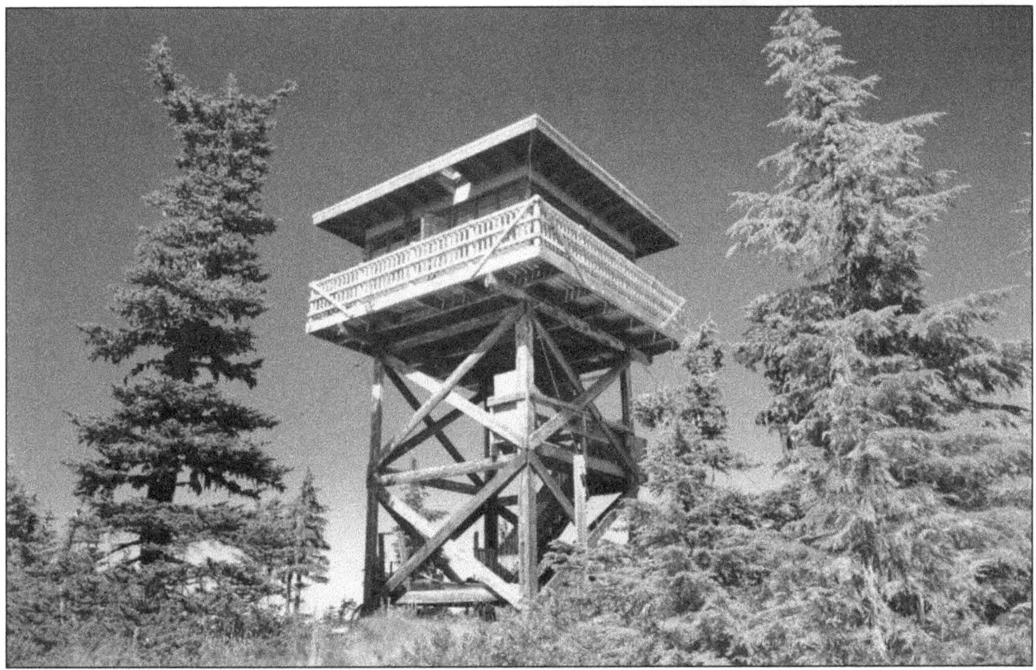

Jim Drury was working the Indian Ridge lookout in 1938 when a lightning storm rolled in. The lookout was struck by lightning, which was witnessed by nearby lookouts who feared Drury had been outside at the time disconnecting the phone. Drury was in fact okay, but the telephone's fuses and wiring were burned. He finally repaired the phone at 4:30 a.m. and called to let everyone know he was okay. Pictured in 2015 is the current tower on Indian Ridge, which is now available for rent.

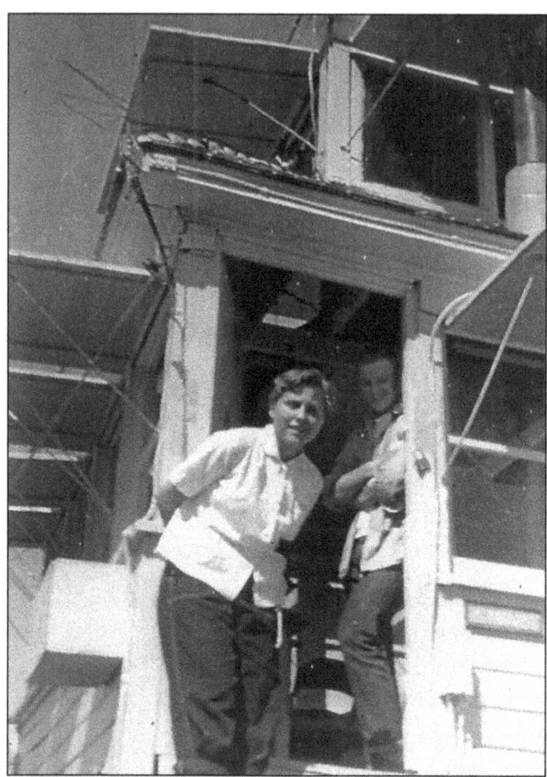

The Fuji Mountain cupola lookout was built in 1928. Patti Luse (left) remembers annual family camping trips at Gold Lake in the 1950s when the whole family would hike up to Fuji Mountain. Luse enjoyed these excursions since it meant getting to visit Larry Aldrich (right), the cute college boy stationed there. An R-6 cabin was built in 1959, but when it ceased to be used, the Forest Service burned it down in 1968. (Patti Luse.)

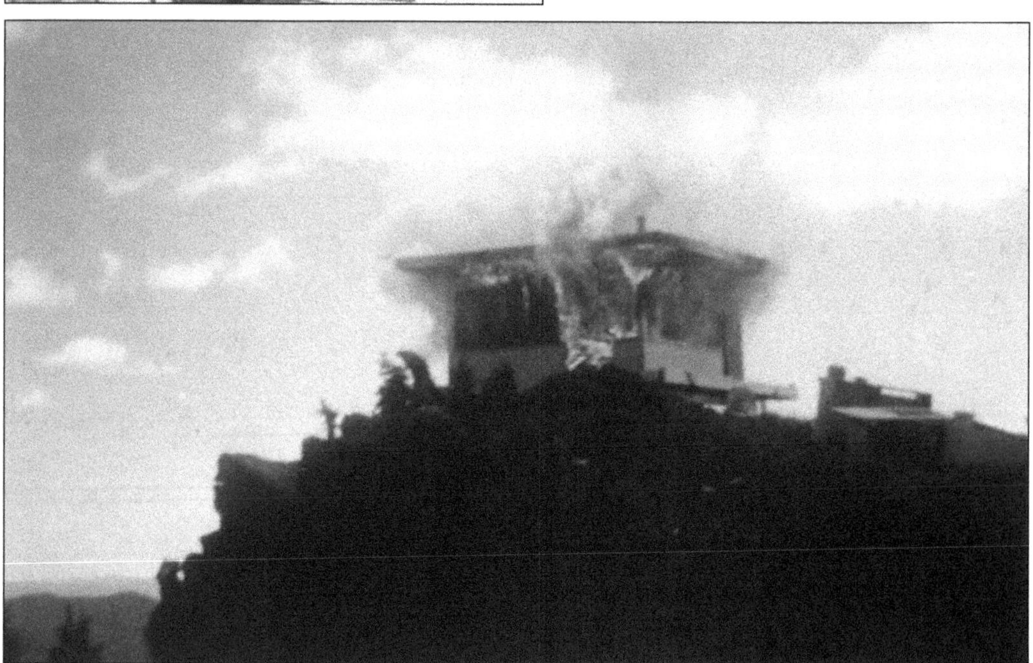

By the 1960s, many lookouts were no longer used, and the Forest Service, worried about liability issues if a visitor was to become injured while exploring an abandoned lookout, removed many of them. Ironically, a favored method of removal was fire, as shown here on Fuji Mountain in 1968. (LCHM.)

Four

SOUTH CASCADES

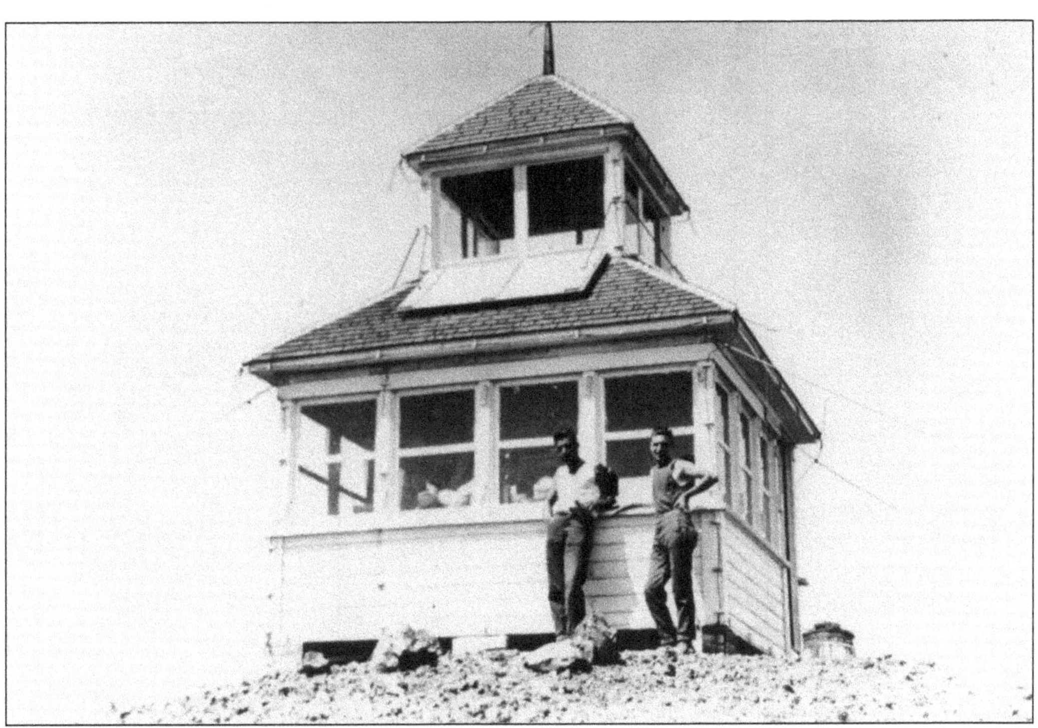

A D-6 cupola cabin was built on Mount Bailey in 1923. Henry Homalac was stationed there one summer when a thunderstorm rolled in. The cabin took a direct lightning hit, burned out the phone, and knocked him out for several hours. When he woke up and discovered his phone was dead, he staggered down the trail until meeting the person who had been sent up to check on him.

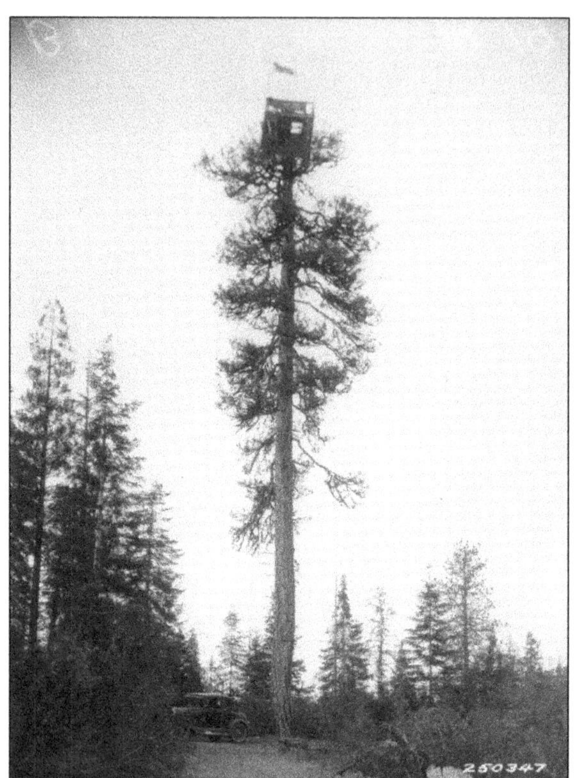

A crow's nest lookout was established on Agency Hill in 1930. It was 106 feet up a tree. Perhaps because of the dizzying height, it was quickly replaced by a proper tower. The 110-foot-tall tower stood for nearly 50 years before its removal in 1979. (National Archives.)

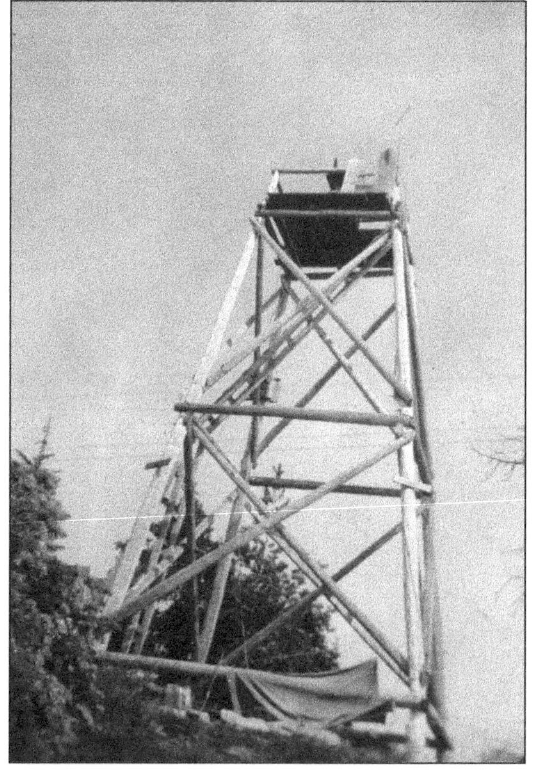

The first lookout on Elephant Mountain was this 20-foot-tall tower with an open platform on top, built around 1928. In the summer of 1930, Harold Dahl was stationed here. The ground cabin where Dahl slept was plagued with noisy wood rats, and one day, he decided to take a shot at one of them. "When I pulled the trigger, the concussion raised every loose particle on the floor and thereabouts and deposited it in many places, including over myself, the bed, cache box, store," said Dahl. But he did manage to kill the rat.

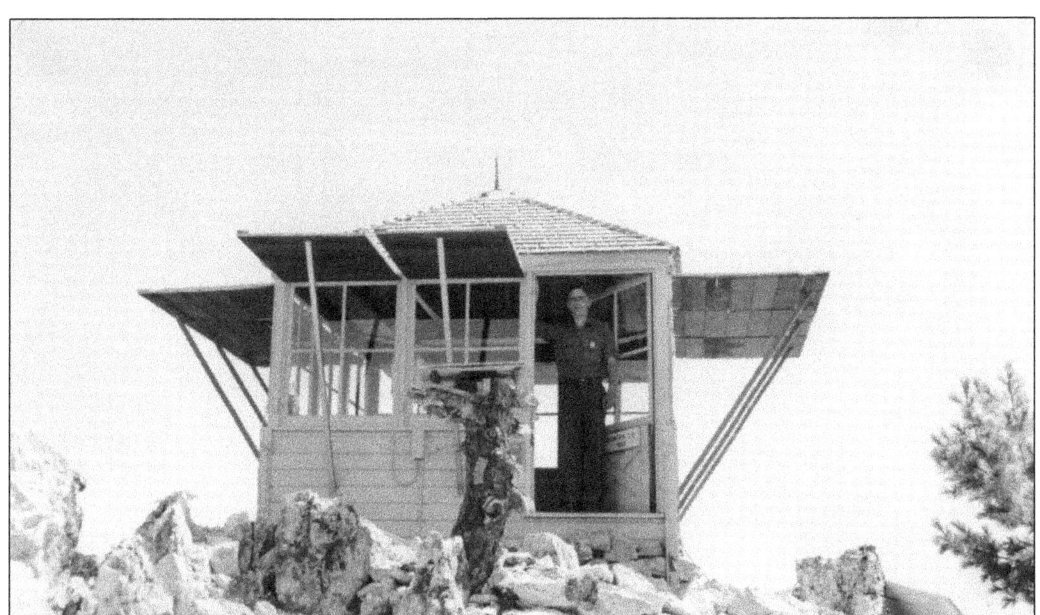

In the early 1930s, Joe Smith—the pilot who flew the Pacific Air Transport mail plane between Portland and Oakland—liked to drop an *Oregonian* newspaper as he flew over Quartz Mountain, figuring that lookout Howard Church would appreciate some news from the outside world. One time, Smith's goggles blew off as he flew over the mountain, and he figured they were lost. But 10 days later, they showed up in the mail with a note: "I found them hanging in the branch of a tree." This 1940 photograph shows the second lookout on Quartz Mountain, which was built in 1933.

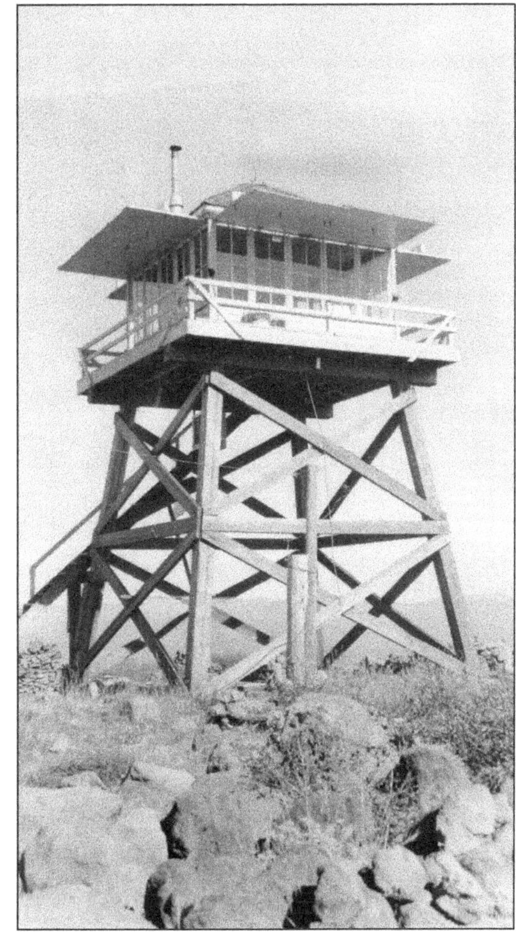

A cupola cabin atop a 15-foot-tall tower was built on Abbott Butte in 1928. The year before, Andy Harvey, George Barrows, Delbert Poole, and a string of nine pack animals took four days to haul the building materials 12 miles along the trail, which in places was buried under four feet of snow. The first lookout was replaced with an L-4 tower (pictured) in 1939. It has been abandoned for decades and now sits inside the Rogue–Umpqua Divide Wilderness.

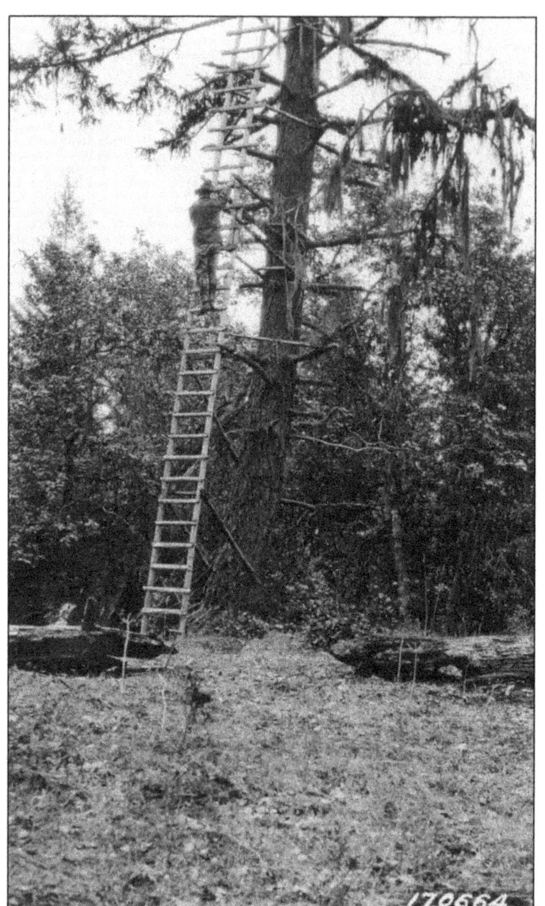

Jack Smith was stationed on Callahan Mountain in 1933. The lookout was a six-by-six-foot crow's nest in a tree, and to access it, Smith had to climb an 83-foot ladder. He wrote that, although he had a few visitors that summer, he "never could get a visitor more than 20 feet up the ladder."

A 16-by-16-foot cupola cabin was built on Calimus Butte in 1920, replaced by another cupola cabin in 1930. Originally operated by the Klamath Indian Agency (then the Oregon Department of Forestry), the lookout is now owned by the Fremont-Winema National Forest.

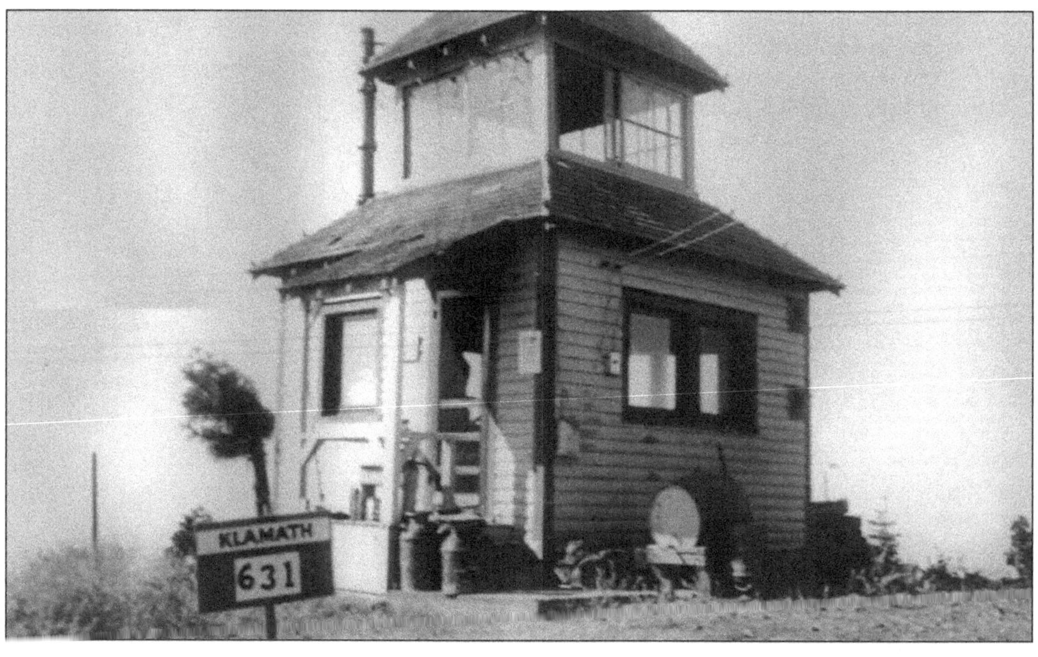

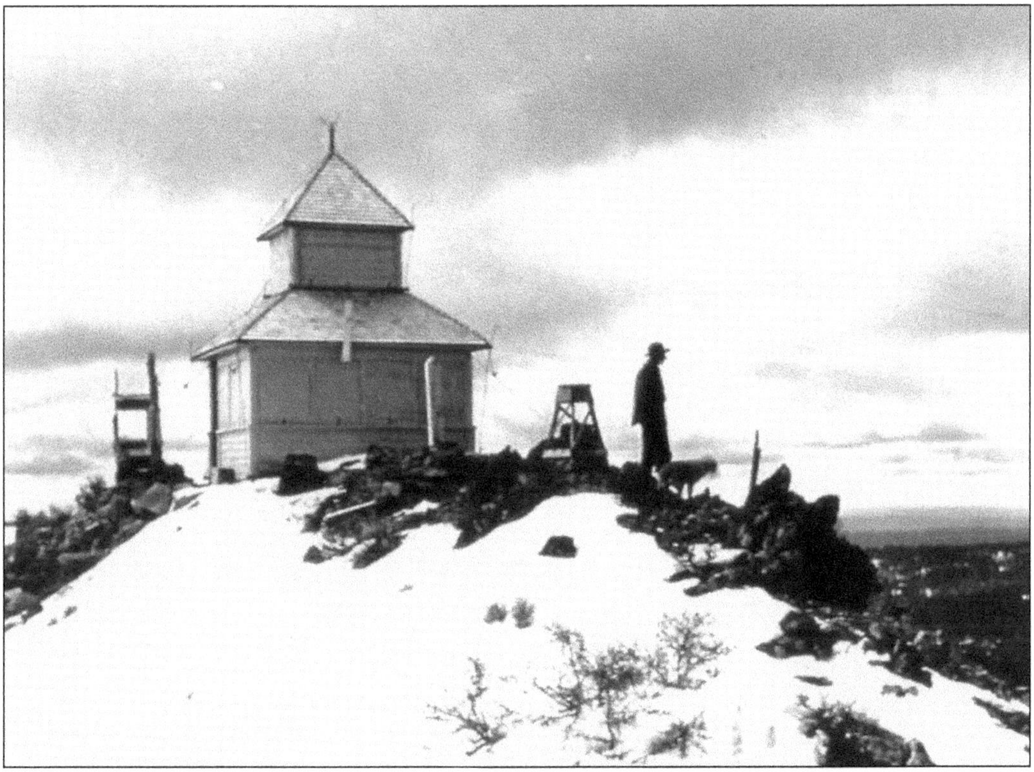

This 1929 photograph shows the first lookout on Dog Mountain. It was replaced by an L-4 lookout in 1947, which was then replaced by a two-story lookout in 1997. This lookout was the first to report the Barry Point Fire in August 2012. The fire lasted three weeks and burned 93,000 acres. Ironically, it burned up the slopes of Dog Mountain, but crews had encased the structure in protective wrapping and dropped fire retardant, and the lookout was saved.

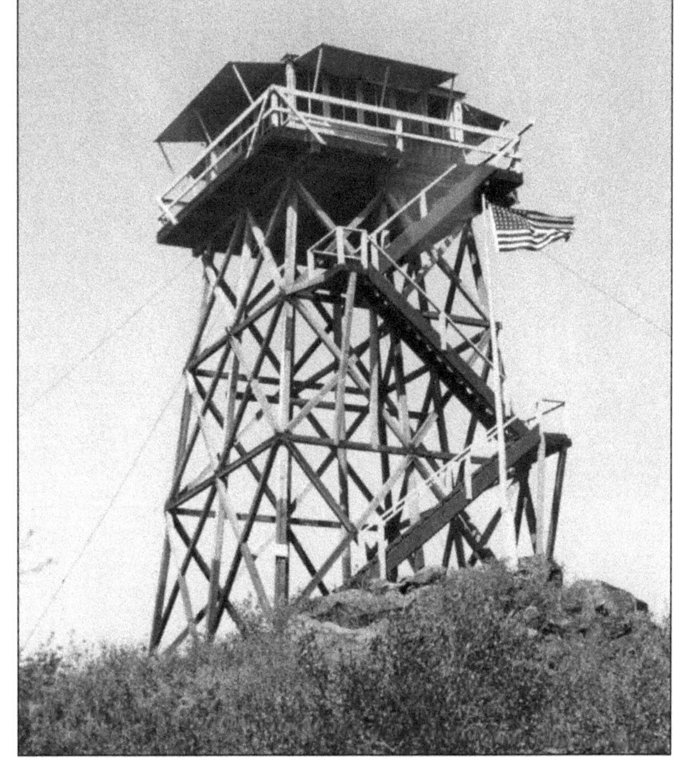

The first lookout on Horse Fly Mountain was a 37-foot-tall L-4 tower built in 1934, pictured in 1945. It was replaced by a 41-foot-tall tower in 1961 and is still staffed every summer. (National Archives.)

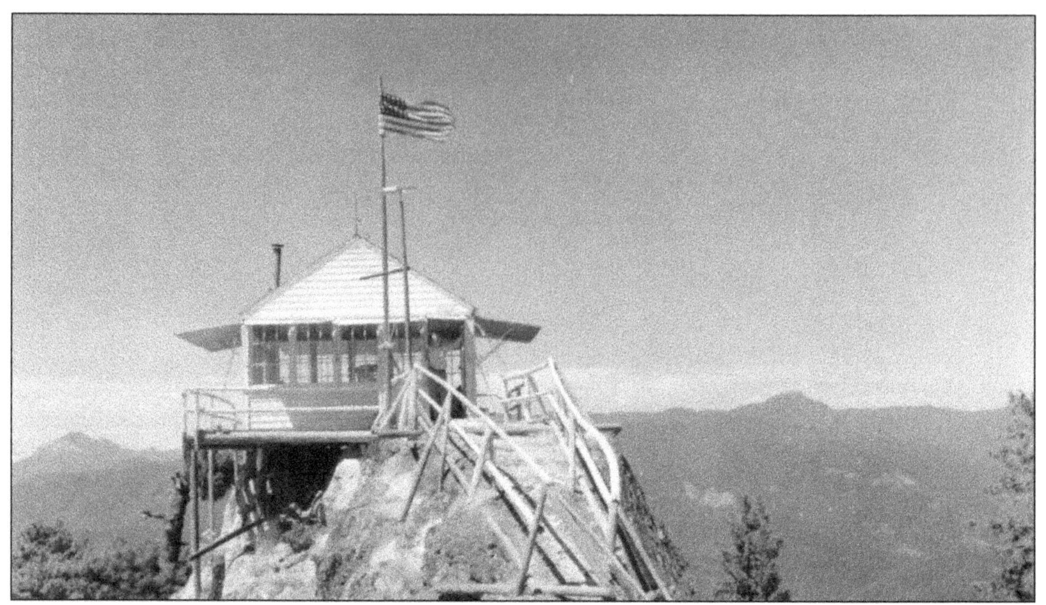

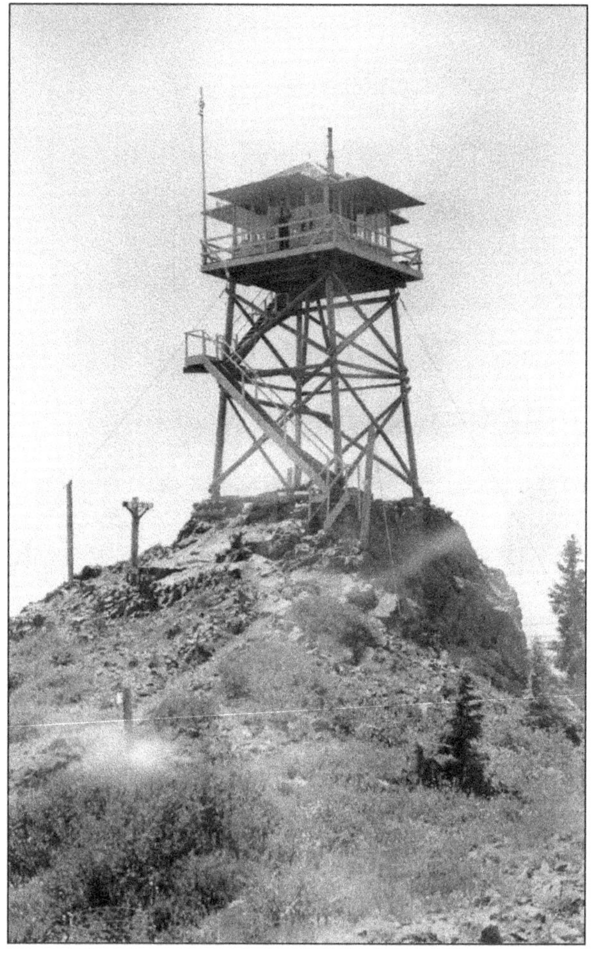

In 1945, Anne Grace Chapple visited her son Phil, who was stationed at Acker Rock, pictured in 1942. She marveled that her son had to clean all 180 panes of glass in the windows every two weeks. The three-mile-long hike Chapple had to undertake to reach the lookout has since been shortened to a half mile, thanks to a road, and the cabin is now available for rent.

The first lookout on Fairview Peak was built in 1912. The second was a 20-foot-tall tower (pictured) that was built in 1936 and replaced in 1972 by a tower that still stands today. Doug Newman staffed Fairview Peak in 1967 and remembered the powerful lightning storms he witnessed. He called them "pyrotechnics that would put any Fourth of July fireworks to shame."

Clair Hogate and Frank Wilson recalled the 1949 construction of the lookout on Mount June. It was October and snowing. "We had to tie ourselves up on the roof when we were shingling to keep the wind from blowing us off the tower," said Wilson. "You could only stay up and work 15 to 20 minutes, and then you'd have to come down and warm up." (LCHM.)

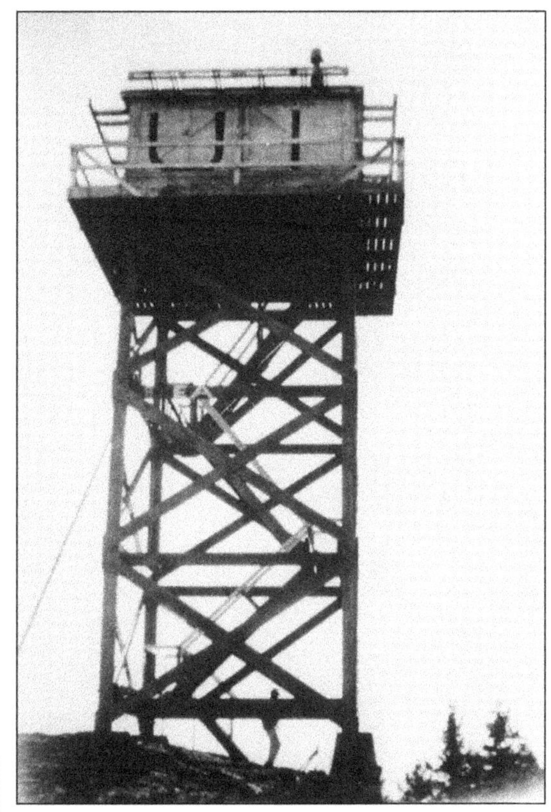

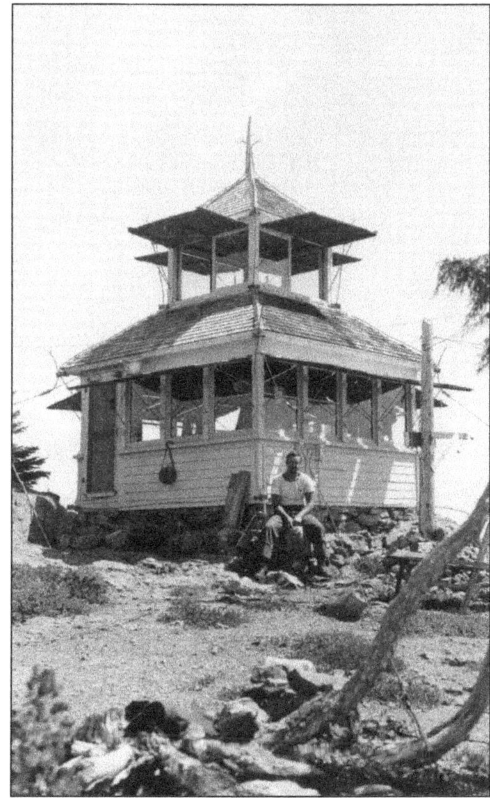

This 1941 photograph shows the first lookout that was constructed on Illahee Rock in 1925. A 20-foot-tall tower replaced it in 1956, and the following summer when Dave Patterson was stationed there, a helicopter crashed while attempting to land on the summit. He found the wreckage below and summoned help for the lone survivor.

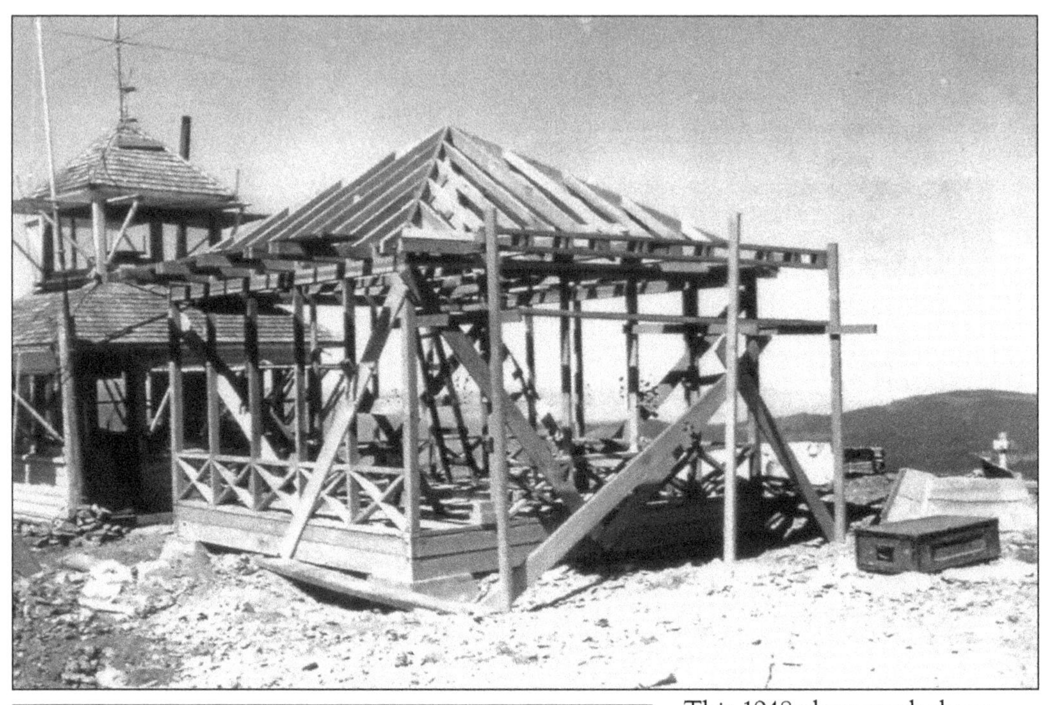

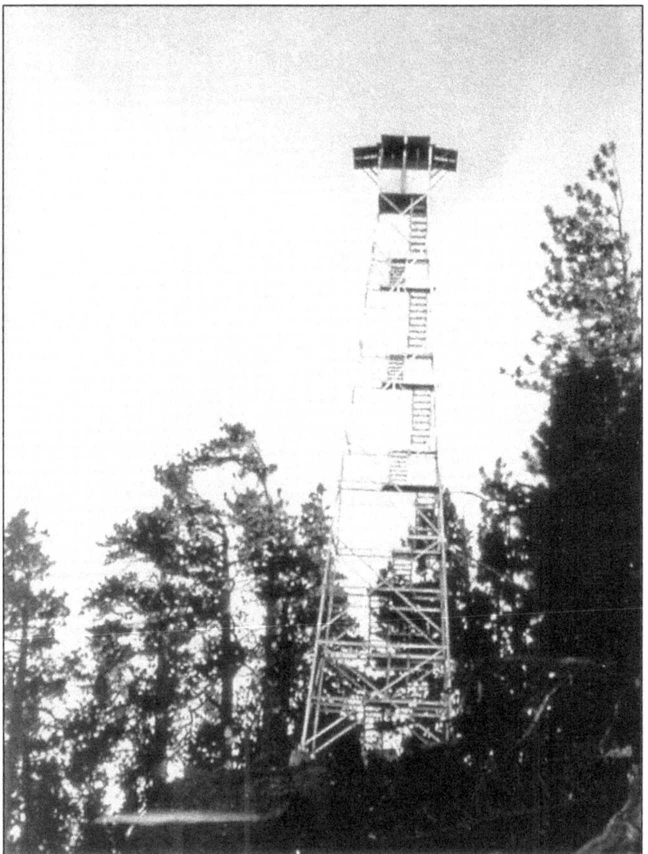

This 1948 photograph shows the new L-4 lookout being constructed next to the old cupola lookout on Drake Peak. It still stands, and although it is no longer staffed, it is part of the rental program. At 8,222 feet, it is the highest Oregon lookout accessible by vehicle, as well as the highest rentable lookout.

A 72-foot-tall tower with ground cabin living quarters was built on Fremont Point in 1936. The cabin was part of the rental program when it burned down in the 2002 Winter Fire. The steel tower was also damaged in the fire, and the Forest Service removed it.

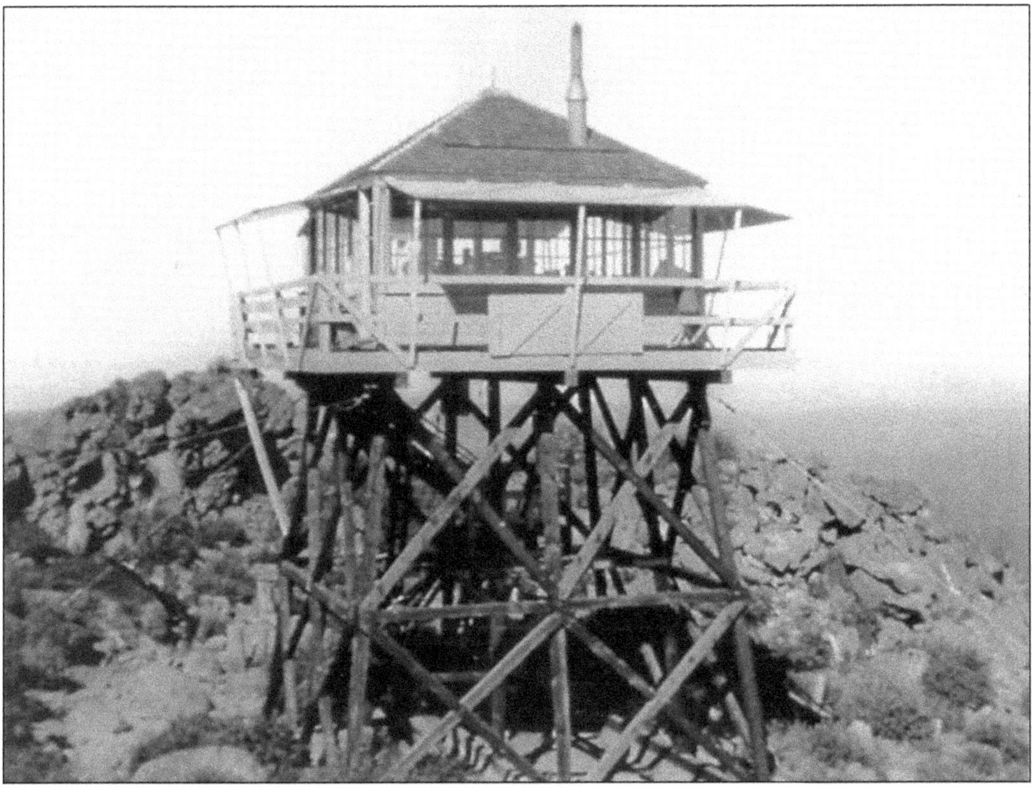

The 16-foot-tall lookout on Rodman Rock was built in 1933. The Silver Lake Ranger District conducted a fire school near Rodman Rock in the 1940s, where men learned methods of firefighting and prevention. At the lookout, they learned how to spot fires and determine their location.

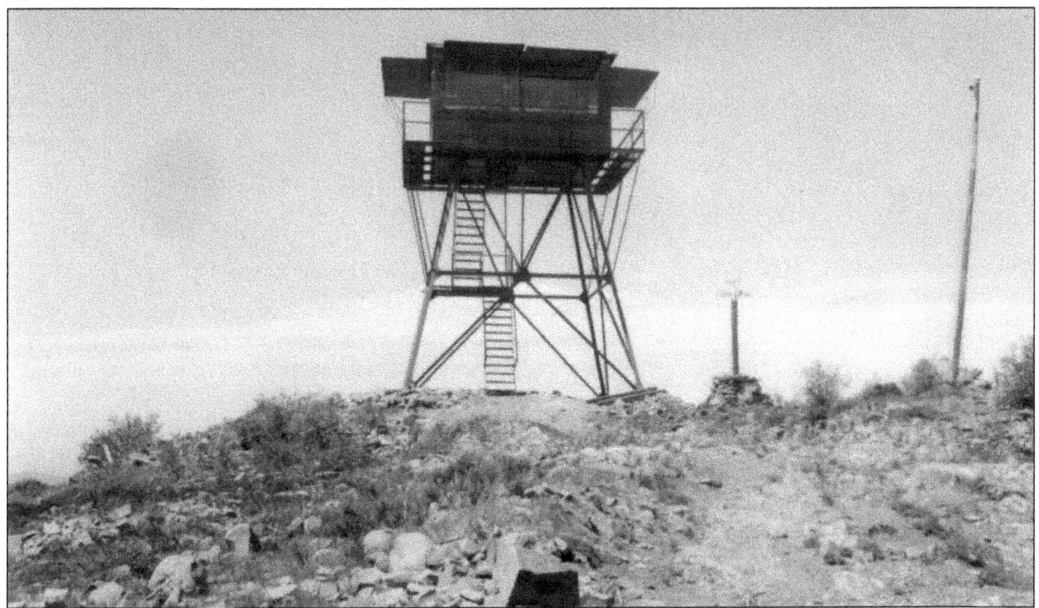

The first lookout on Chase Mountain was built in 1928. This 20-foot-tall steel tower replaced it in 1951. It is staffed every summer and operated by the Klamath Forest Protective Association. (ODF.)

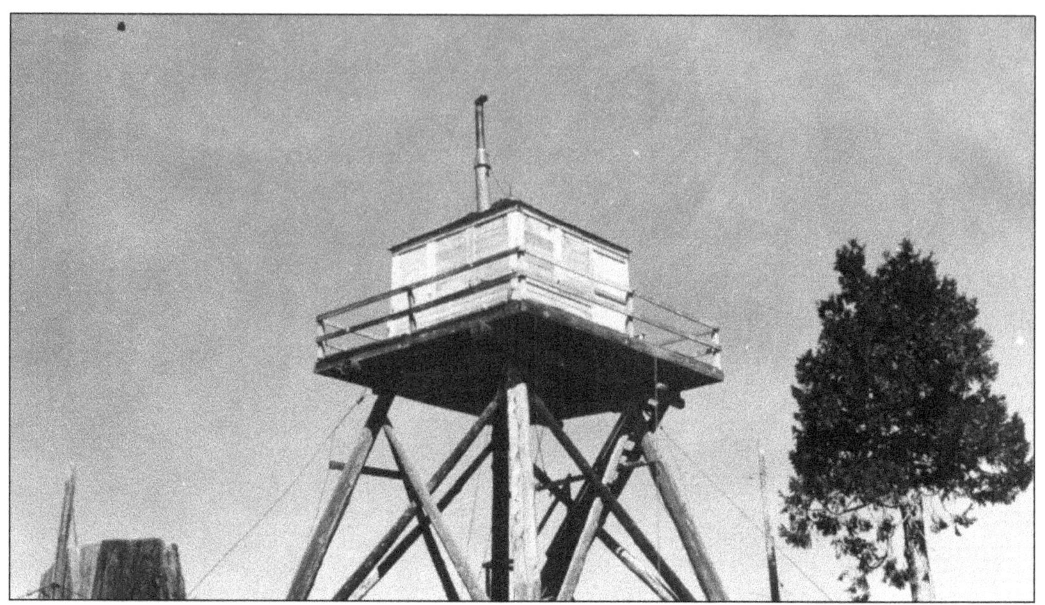

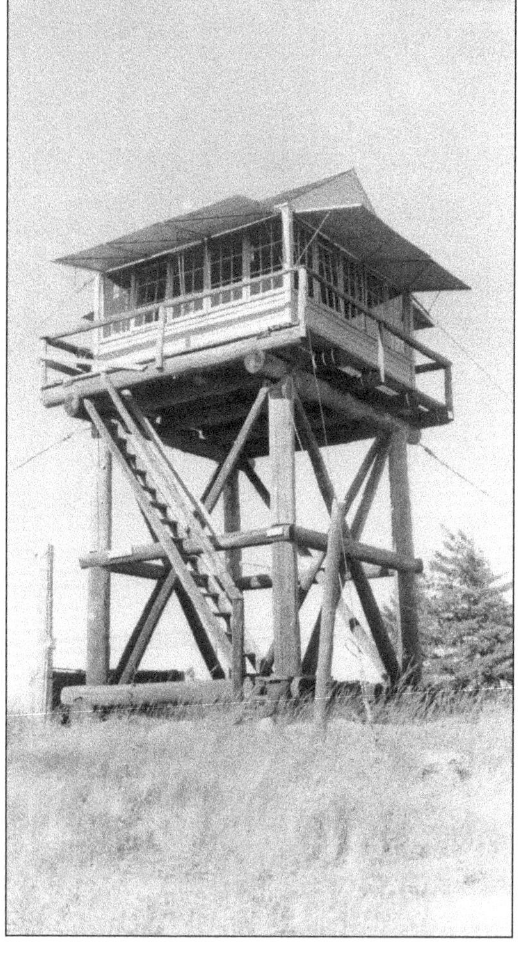

The Civilian Conservation Corps built this 30-foot-tall tower in 1934 at a place known as Anderson Camp, on Anderson Mountain, which had once been the location of a sheepherder's camp that was named after Frank Anderson. Clinton Wynn was stationed here in 1938 and 1939 and was paid $100 per month, with $5 deducted to pay "rent" for the lookout. The tower was abandoned in the 1950s and later removed.

Shortly after this photograph was taken of the Butler Butte lookout in July 1942, a ground cabin was constructed to house the couple who would be stationed here as part of the Aircraft Warning Service. Clarence and Laura Hartley spent 13 months living on Butler Butte, monitoring the surrounding airspace for enemy planes and balloons. Every six weeks in the winter, Forest Service workers snowshoed in with fresh supplies and mail.

This 30-foot-tall tower was built by the Civilian Conservation Corps on 5,048-foot Holland Point in 1933. The mountain and the meadows directly east were named after Hugh Holland, a settler who had a cabin in the area that he used for hunting and trapping. The tower has since been removed.

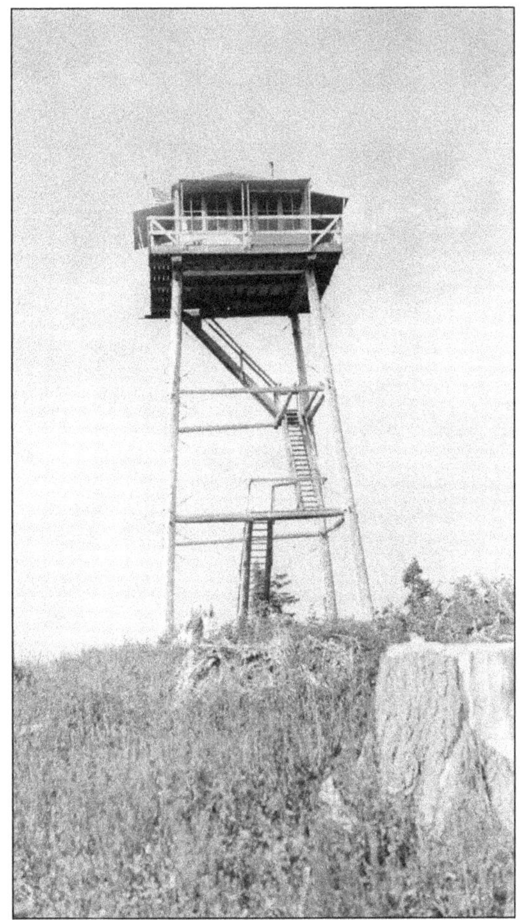

Ruth McMillan, who was stationed at Devils Knob in 1920, had a terrible time with mice in the lookout cabin. One night, she caught 12 mice. This 1942 photograph shows the third and final lookout that sat on Devils Knob from 1936 to 1962.

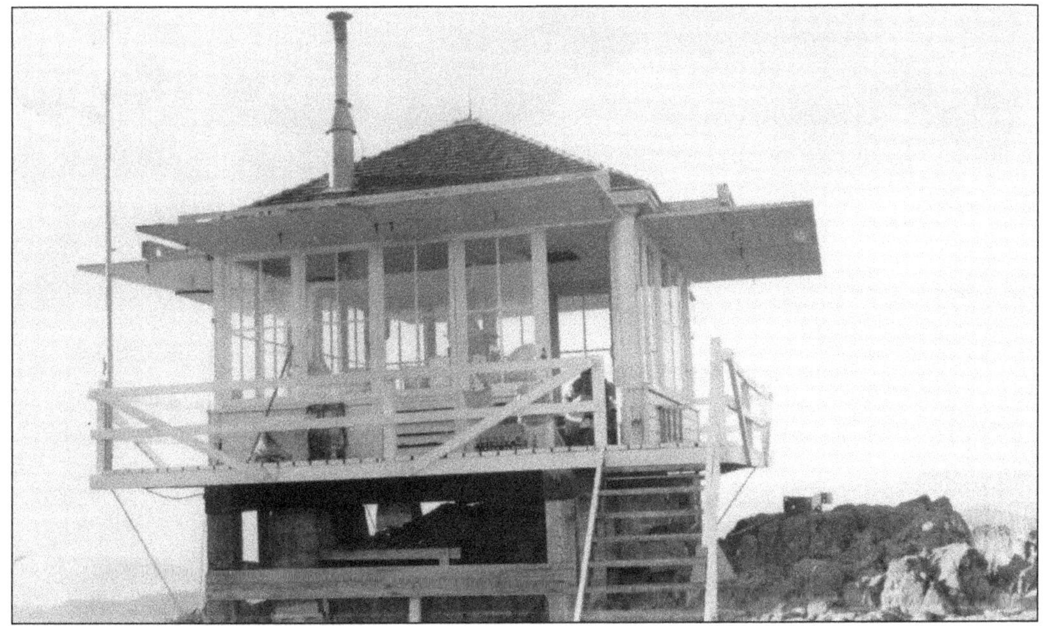

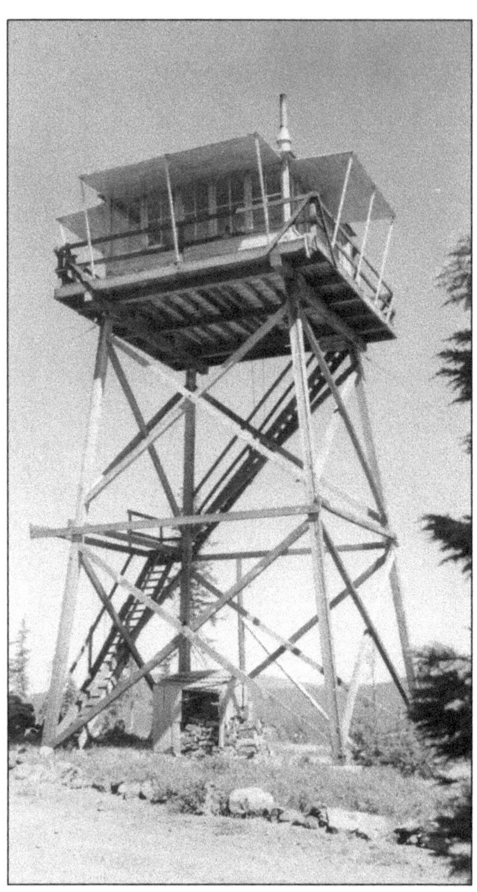

The tower pictured here was constructed on Cinnamon Butte in 1934. It was deemed unsafe at the end of the 1975 season, so when Gary Deboi was stationed on the butte the following summer, he lived in a trailer and had to hike around to different points on the butte in order to make his observations. The old tower was knocked down in July and replaced by a tower that was moved over from Buster Butte. It still stands and is staffed in the summer.

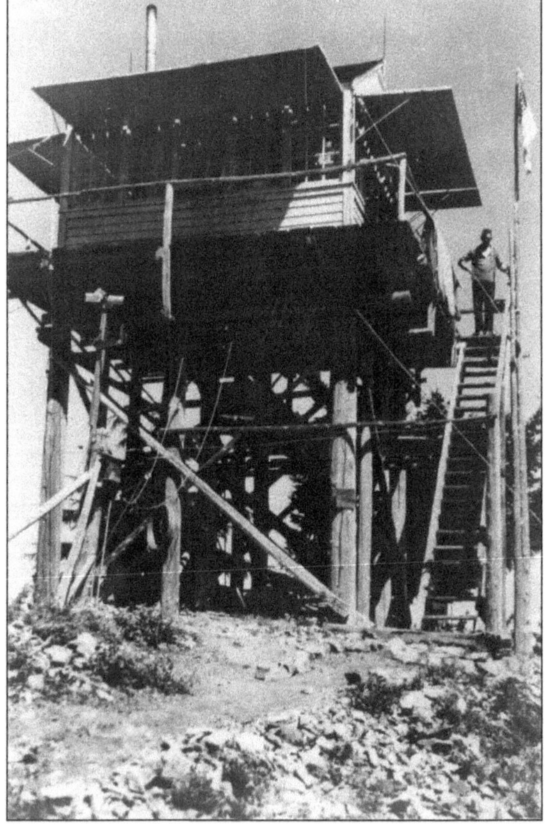

The lookout on Silica Mountain, pictured in 1940, was built in 1931. One summer, the lookout reported seeing a "big kitty" around the summit. When the fire control officer visited the station and walked out on the catwalk, he saw a large cougar lounging in the sun 50 yards down the mountain. "Big kitty" indeed.

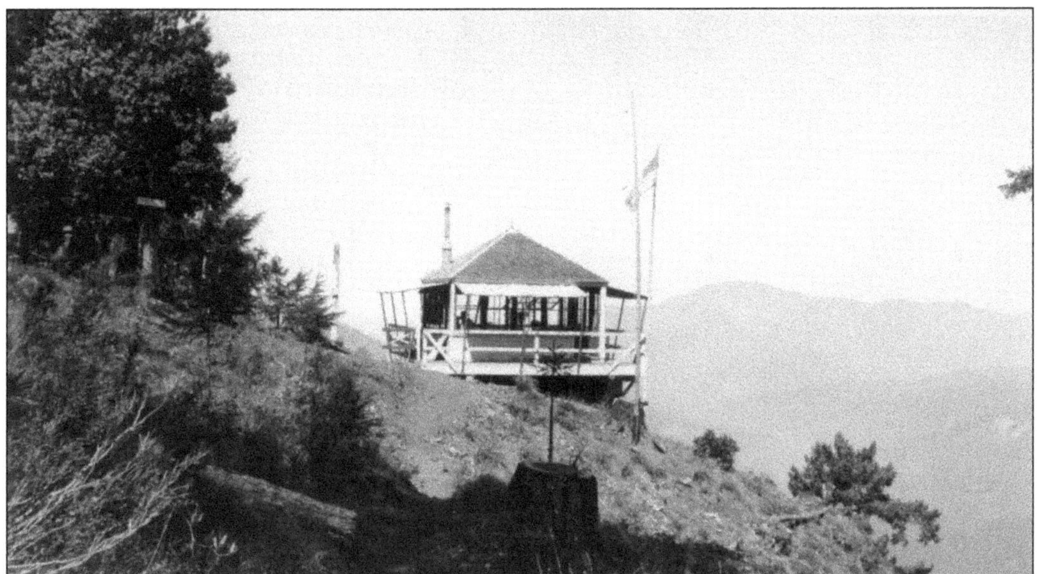

This L-5 cab was built on Shivigny Mountain by the Civilian Conservation Corps in 1933. The mountain was named after Emile Shivigny, who settled in the area around 1875. The original cabin was replaced by a 54-foot-tall steel lookout tower in 1948, which was moved to Baughman Point in the 1970s.

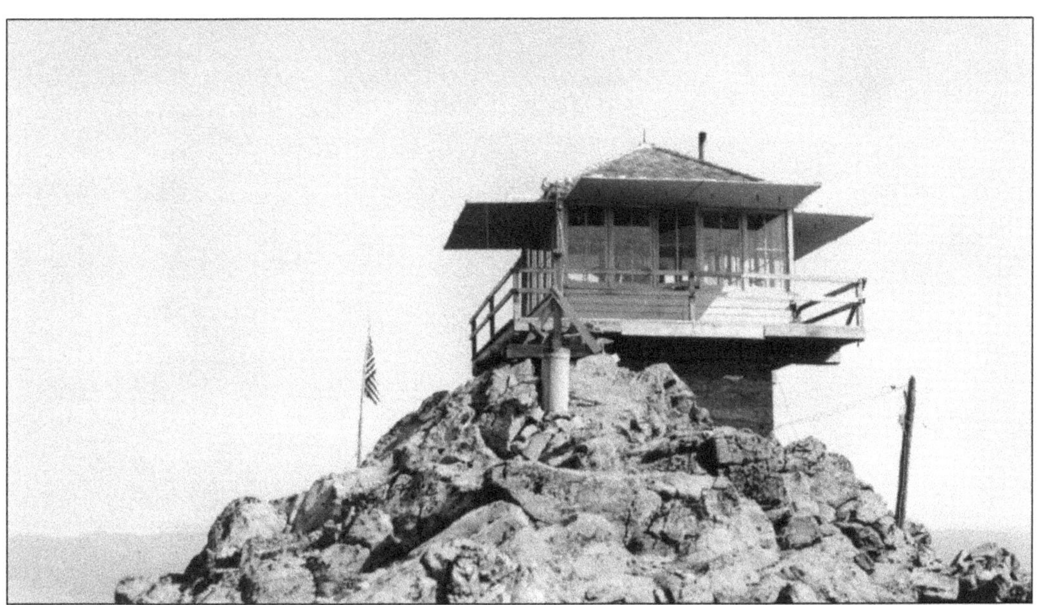

The first lookout on Black Rock in the Umpqua National Forest was built in 1911. This L-4 lookout was built in 1938. When Andy Coleman was working here in the summer of 1961, he made the mistake of trying to refill his gas lantern while it was still hot. The flash could be seen 16 miles away by the lookout on Tallow Butte. Fortunately, the lookout did not burn down.

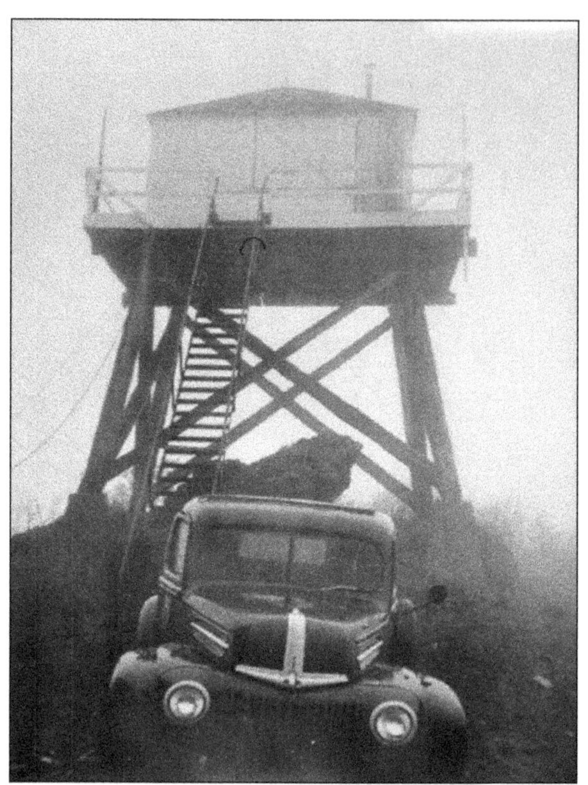

During the summer of 1953, Jack and Lorraine Holmes worked as lookouts at Buck Peak. They brought along their seven-year-old daughter, Sharon. Jack was paid $210 a month as lookout, and Lorraine was paid $50 to serve as auxiliary lookout. Everyone bathed using a metal tub in the middle of the cabin, using as little water as possible since Jack had to drive to a nearby spring to retrieve it. (ODF.)

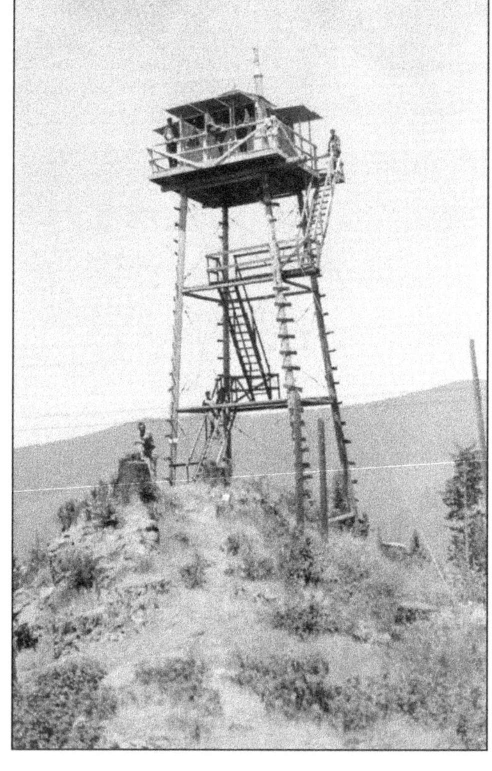

The 40-foot-tall Rujada Point lookout was built on Rose Hill in 1934. One theory of the origin of the name "Rujada" is that it was taken from a telegraph code book and meant "a considerable body of standing timber is available." A lumber camp east of Cottage Grove also carried the name.

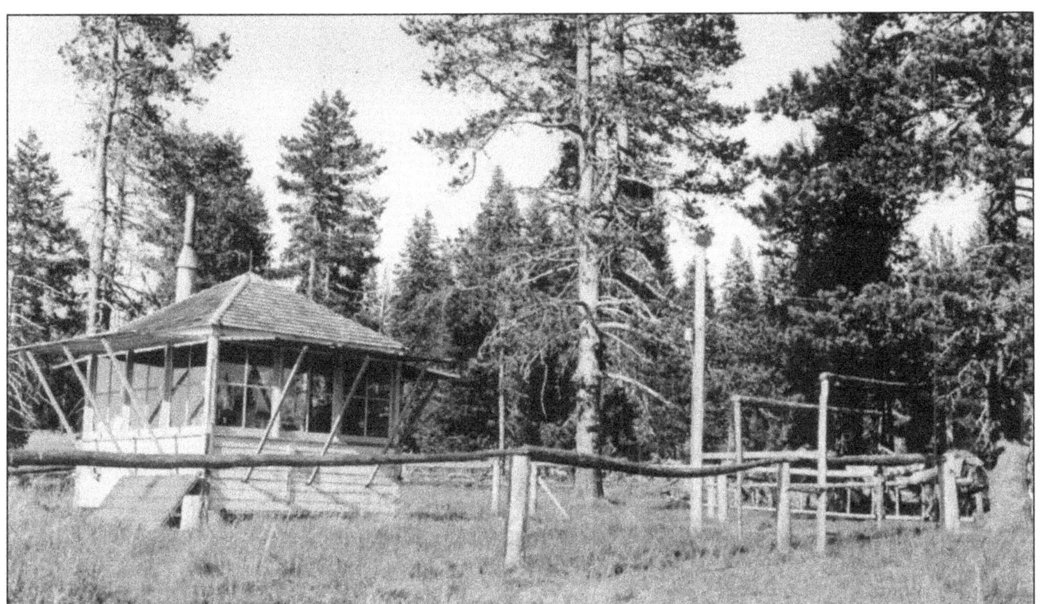

Most lookouts were located atop mountains and buttes to maximize visibility; however, this lookout was built in 1933 in a place called Skookum Prairie in the Umpqua National Forest. The prairie was below Garwood Butte, which also had a lookout starting in 1942.

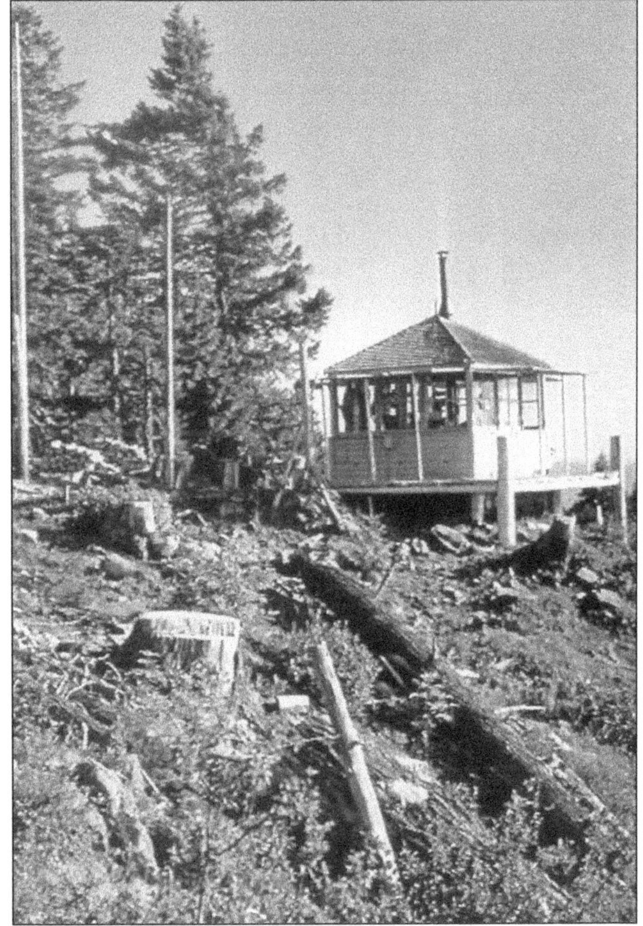

An L-5 lookout cabin was built on Reynolds Ridge in 1933, and Bill Noland was stationed there in 1946. He got permission to cut down a tree that was obstructing his view and to use the wood to build a catwalk for the lookout. This photograph shows the results of his handiwork. He remembers making oatmeal cookies and enjoying them so much that he ate them for breakfast, lunch, and dinner until they were gone. The lookout was removed in 1963. (Bill Noland.)

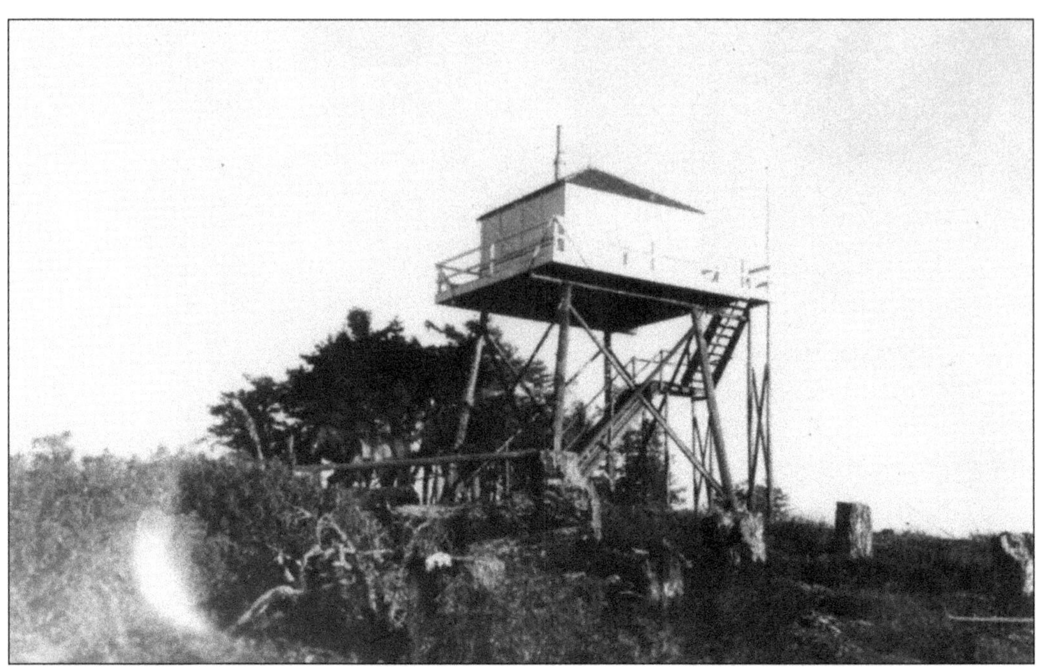

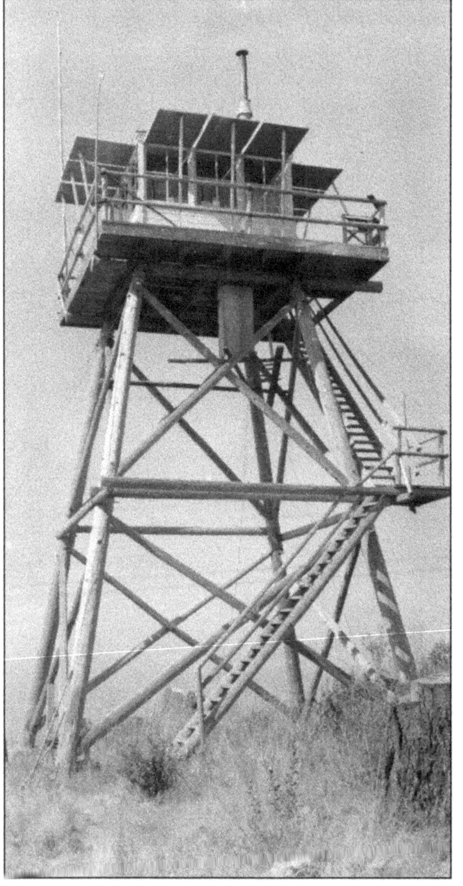

This 30-foot-tall tower was built on Taft Mountain in the Umpqua National Forest in 1933. Before World War I, a huge tree about a half mile from the summit was widely known as the Taft Tree, so named by District Ranger Oscar Houser in honor of the president. The mountain was named after the tree.

Pickett Butte was named for William T. Pickett, a homesteader who settled at nearby Pickett Prairie. In 1934, a 25-foot-tall tower was built on Pickett Butte and replaced by a 41-foot-tall tower in 1940. It still stands and is available for rent when not being staffed during the fire season.

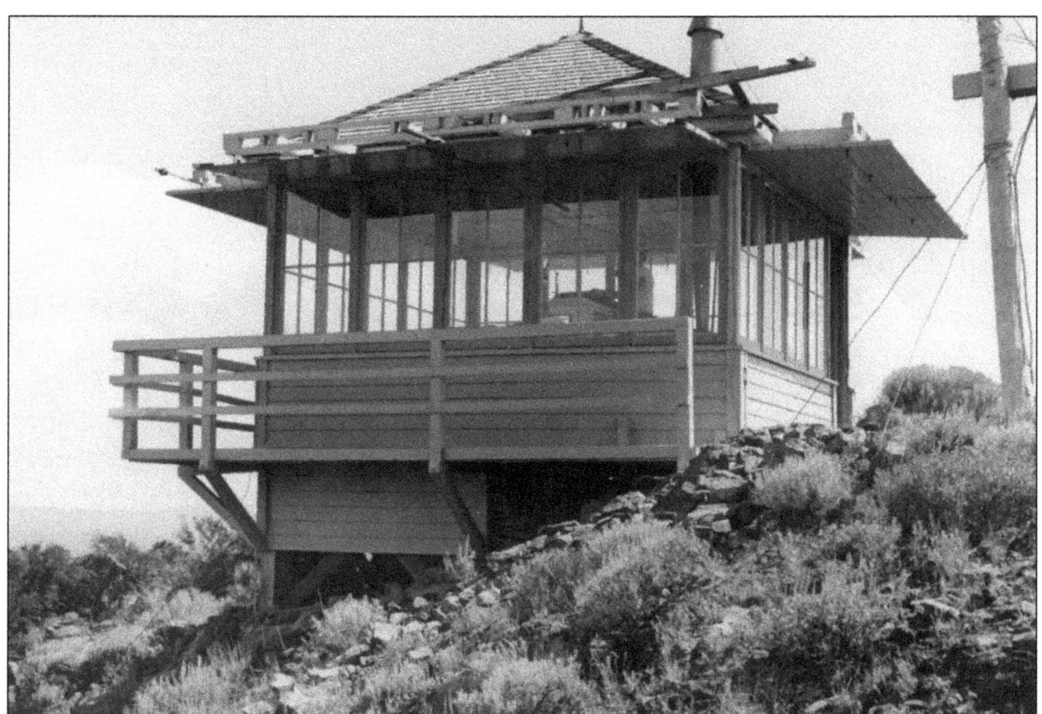

This 1945 photograph shows the second lookout that was built on Hager Mountain in 1936, replacing an earlier 1912 octagonal cabin. An R-6 cabin was built in 1967 and still stands today. It is staffed in the summer and is available to rent the rest of the year. (National Archives.)

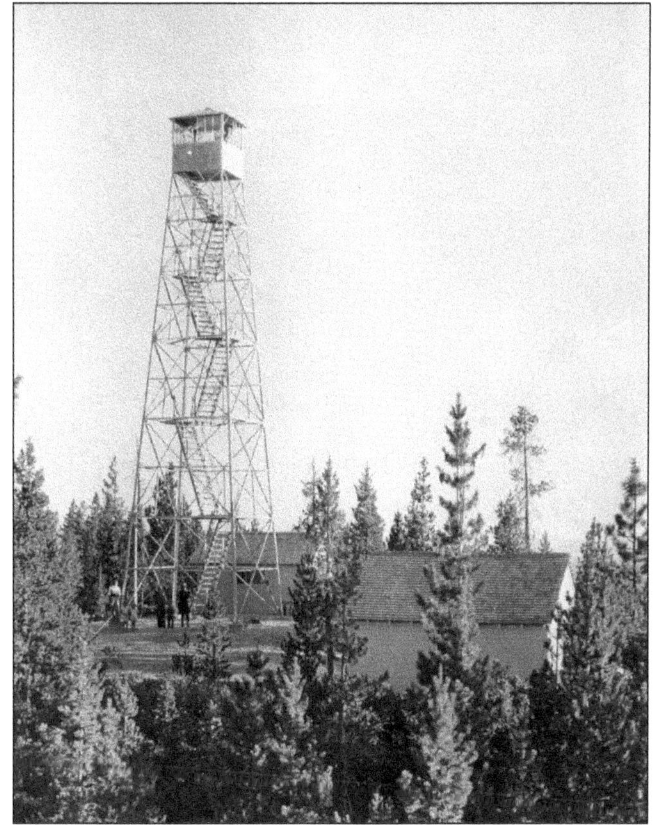

The Forest Service had to battle Mother Nature when erecting the 60-foot-tall steel tower on Skookum Butte in June 1933. With snow on the ground and some areas flooded with snowmelt, trucks had a hard time delivering materials to the summit. The lookout stood for three decades before being removed in 1965. (National Archives.)

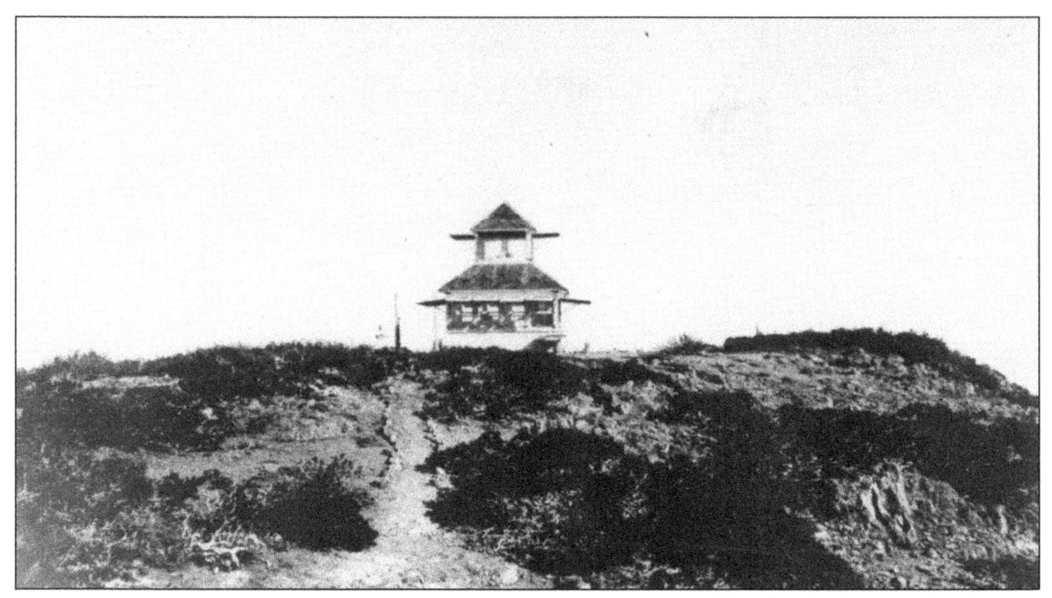

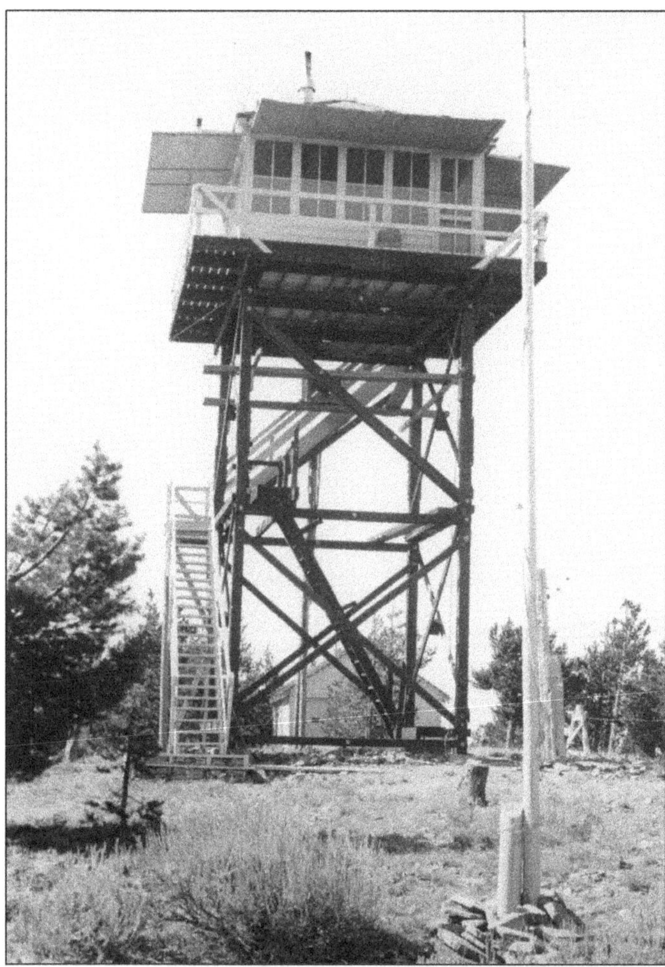

The first lookout on Red Butte was built in 1930 and was later replaced by a new tower in 1953, which now stands abandoned. Howard Church was stationed here in 1935 and received regular deliveries of newspapers from Seattle and San Francisco, dropped by pilots flying over the butte. He sometimes received candy, cigarettes, gum, and magazines as well.

The second lookout on Bald Mountain, pictured in 1945, was built in 1941. Gene and Ruth Groth were stationed at the Bald Mountain lookout in April 1945 when they spotted a strange-looking balloon hanging from some pines and reported it. The balloon turned out to have incendiary bombs attached to it and was one of about 9,300 balloons launched by Japan in hopes of setting fire to the forests of the West and dampening American morale. (National Archives.)

This lookout was built on Rattlesnake Mountain in 1933. In 1937, when Ranger Harold Bowerman was leading a pack train in the vicinity, he thought he smelled something unusual. He later alerted the lookout, who hiked down to investigate and discovered a smoldering stump from an earlier lightning strike.

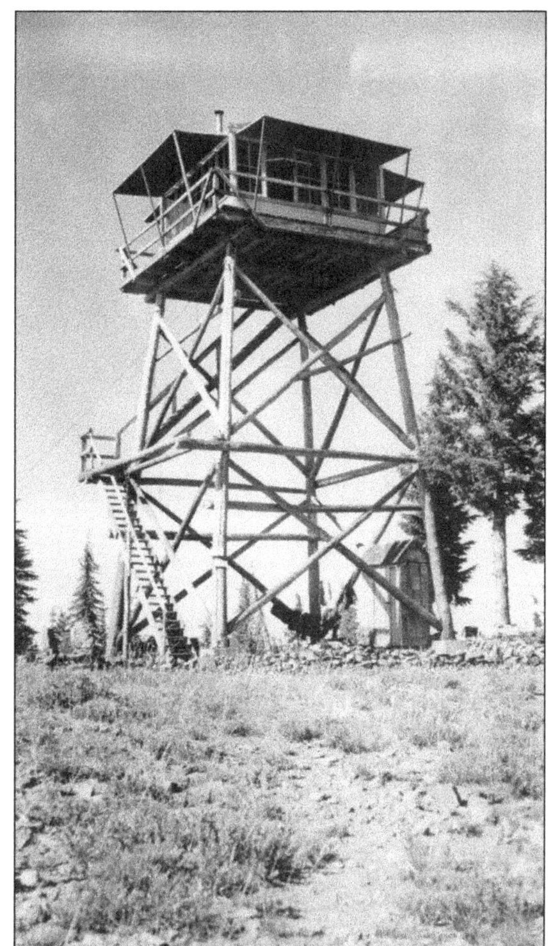

District Ranger Fred Asam visited White Rock in February 1931 for an inspection. Although all looked well from the outside, he discovered that lightning traveling over the line wire had caused some destruction inside. In addition, 14 downstairs windows were broken, and all of the cupola windows were shattered.

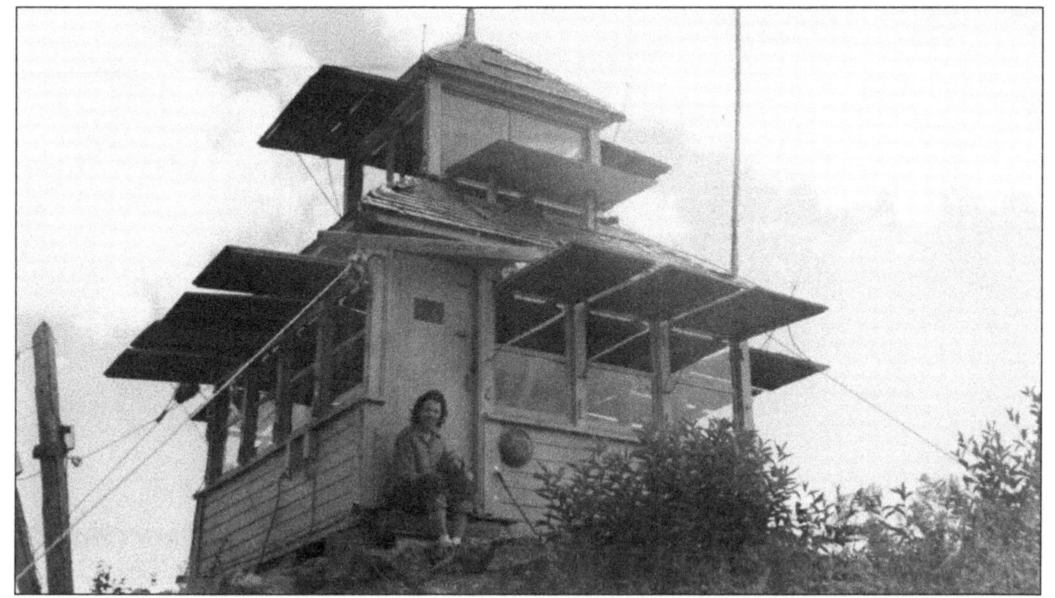

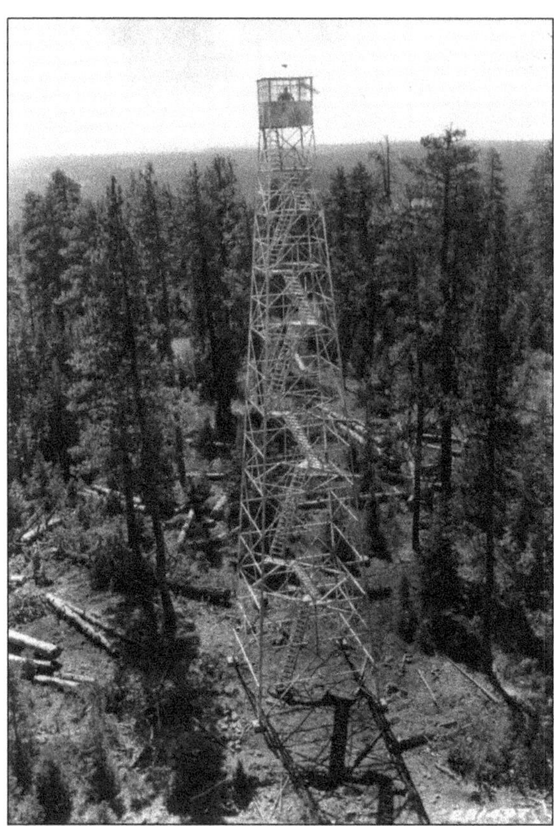

Civilian Conservation Corps men from Camp Ingram built a 101-foot-tall steel tower on Strawberry Butte in 1933. By the late 1960s the tower was no longer regularly used, and it was torn down in 1973.

A crow's nest was established on Red Mountain around 1920, followed by a cupola lookout in 1928. In 1985, the Forest Service moved it off the mountain and relocated the building to the Tiller Ranger Station. It was refurbished the following year and remains on display for visitors.

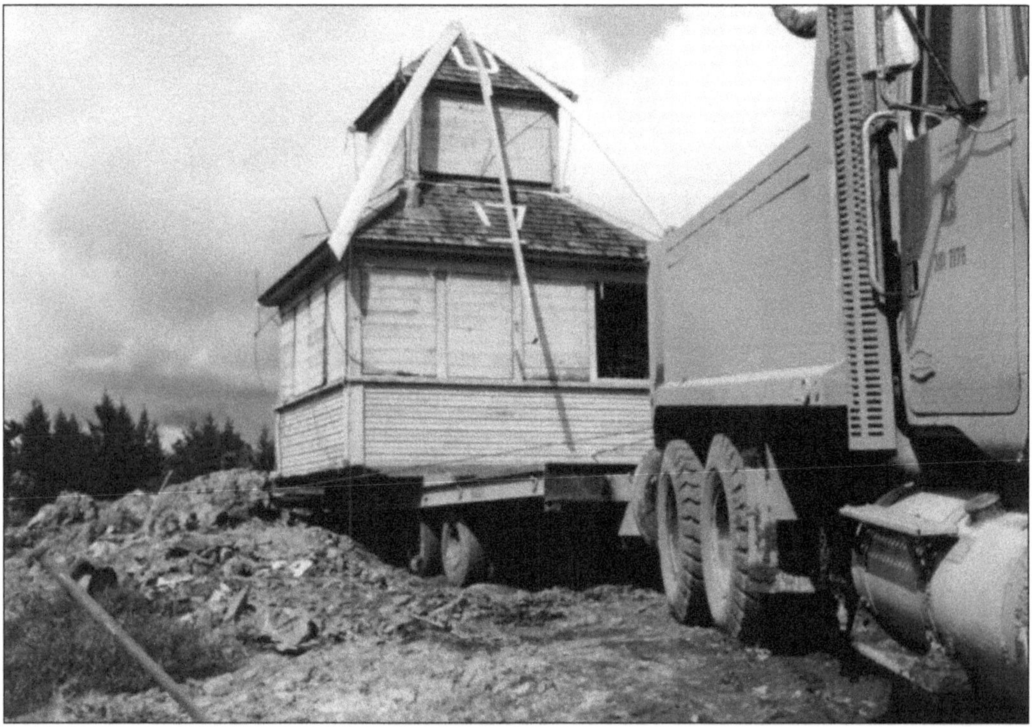

Five

Southwest Oregon

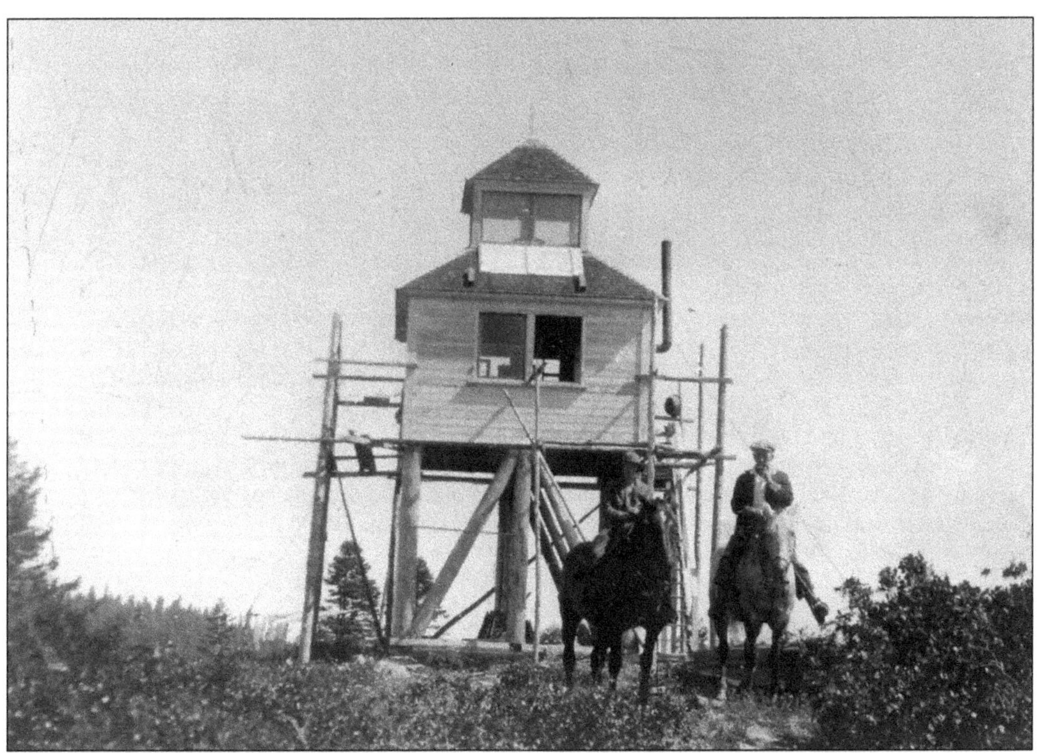

In 1914, a crow's nest was installed 80 feet up a tree at Whetstone Point on Bald Mountain. A D-6 cupola cabin was built in 1920 and was later elevated onto a platform. The lookout was burned down for removal in 1958. (National Archives.)

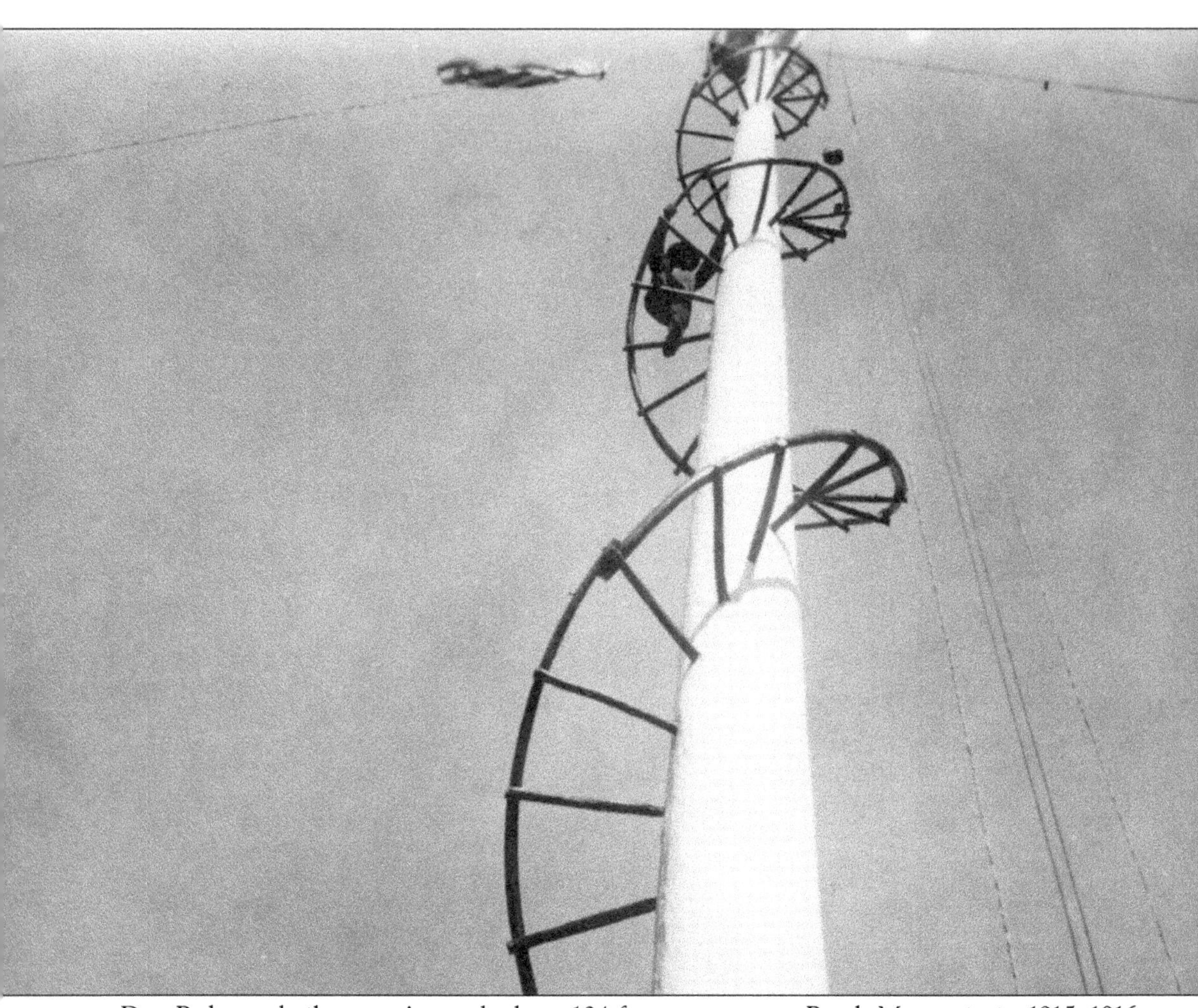

Dan Pederson built a crow's nest lookout 104 feet up a tree on Brush Mountain in 1915–1916. Access to the top was via a spiral ladder consisting of yew pegs driven into the trunk of the tree. During construction, Pederson sat on the highest peg while drilling the hole for the next one up, and did not use any kind of safety harness. Forest examiner Harold Foster noted, "it requires a steady head not possessed by everyone to climb this ladder." Having worked on sailing ships before, Pederson had no fear climbing up and down the ladder. He found the climbing too slow, though, so he rigged up an "elevator" consisting of a cable and two buckets, one of which was filled with rocks to counterbalance his weight. A stone cabin was built in 1918 as living quarters. The lookout was abandoned in the 1930s. (National Archives.)

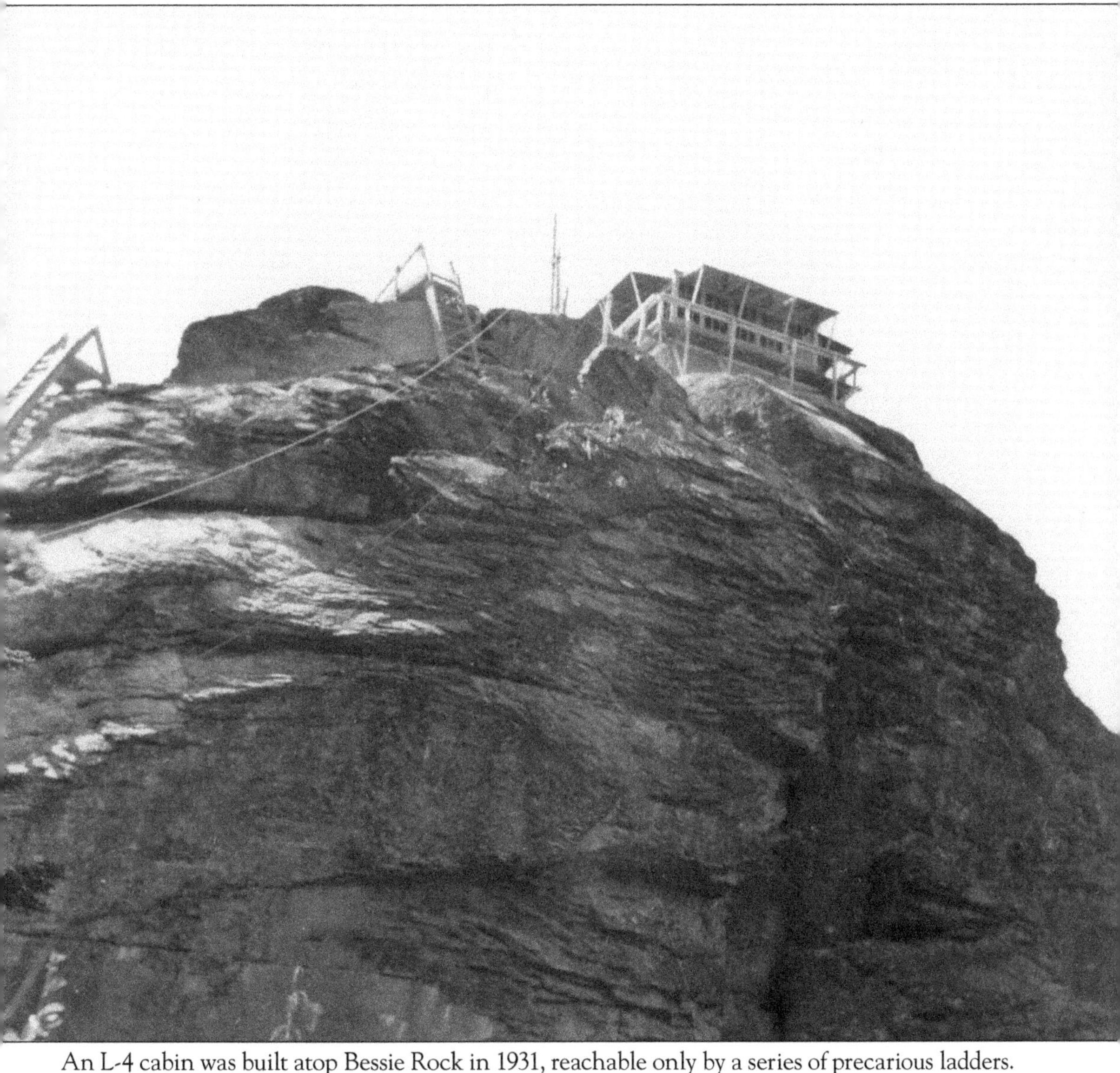

An L-4 cabin was built atop Bessie Rock in 1931, reachable only by a series of precarious ladders. A newly hired lookout was stationed at Bessie Rock one summer when a low-flying jet passed by, and the resulting noise broke all the cabin's windows. The man tried to quit, but his supervisor talked him into staying. Soon thereafter, the cabin was hit by lightning, and the lightning rods melted. His supervisor later discovered that the frightened man had abandoned his post and hightailed it out of there. The lookout was removed in the 1960s. (National Archives.)

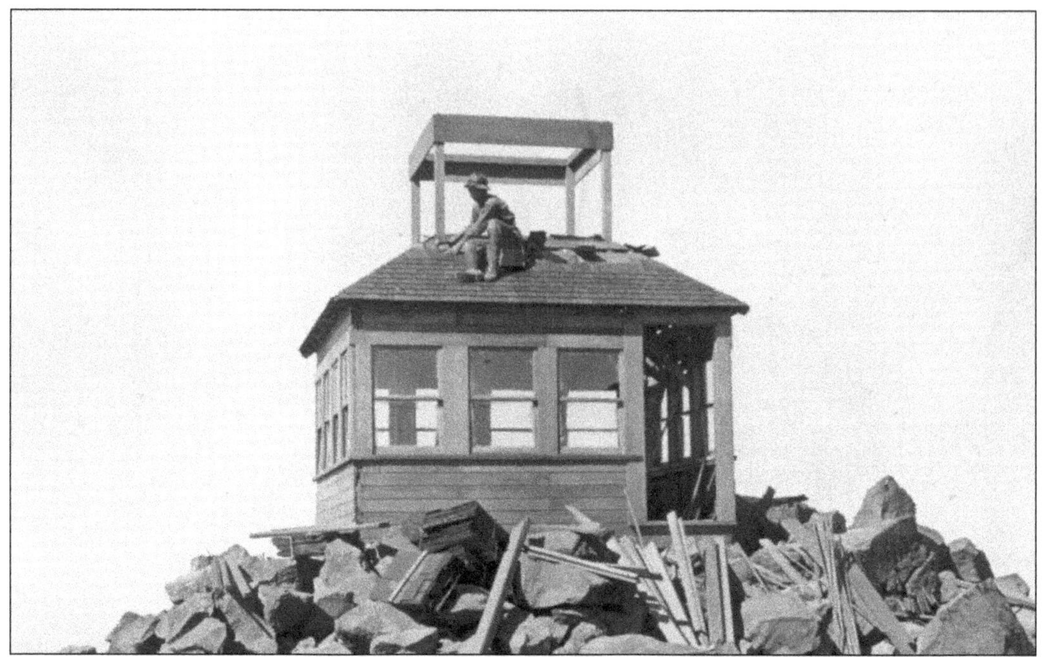

This cupola lookout was built on Mount McLoughlin in 1917. By 1931, the lookout had become unstable, and a new foundation was built; however, five years later, the lookout was deemed too high to be effective and was abandoned, later to be removed in 1955. At 9,495 feet, this is the second-highest lookout site in Oregon. Mount Hood is the highest. (National Archives.)

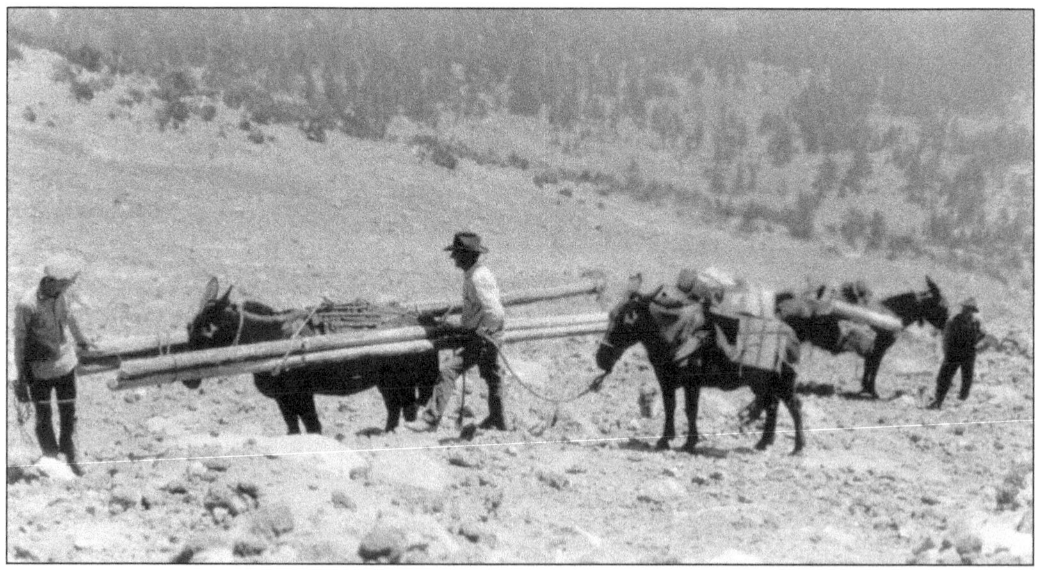

Mules carry supplies and poles up to the Mount McLouglin lookout. During the first few decades of the 20th century, pack animals were often the only way to move supplies around the forested mountains of Oregon. Even once the road-building boom began in the 1930s and 1940s, many lookout sites were only reachable by trail, not road.

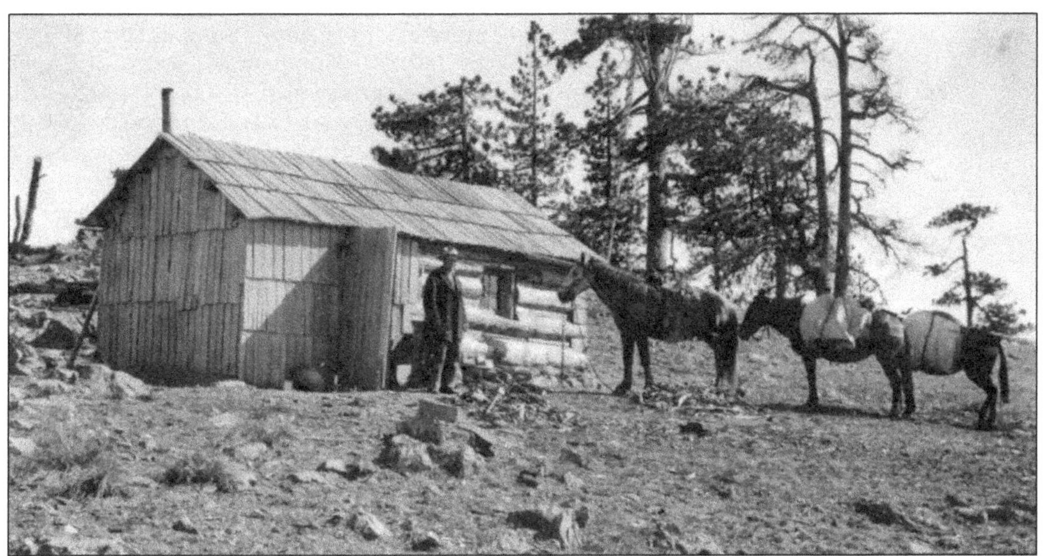

A pack string stands outside the living quarters on Yellowjacket Mountain in this 1917 photograph. The crow's nest lookout can be seen in the background. When the nearby lookout on Dutchman Peak was built in 1927, this site was abandoned. (National Archives.)

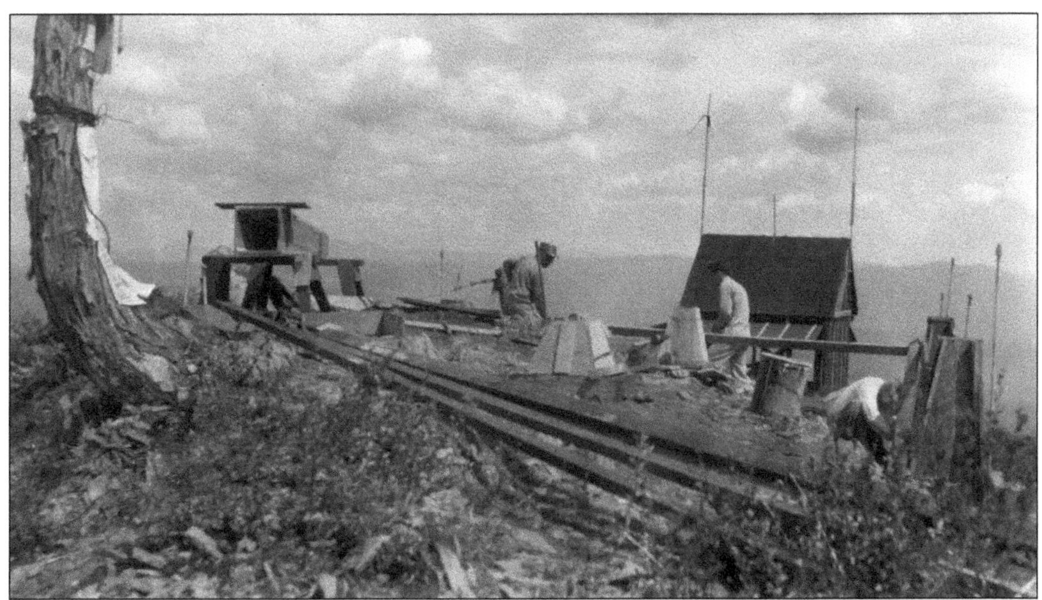

Men construct the Halls Point lookout in 1956, which replaced the nearby Whetstone Point lookout. Constructing a building on a mountaintop was not an easy business, especially if there was no road. Concrete, lumber, nails, tools, window glass, and more all had to be transported to the site, often by mules. Water had to be hauled in from the nearest spring or lake, which could be several miles away. (National Archives.)

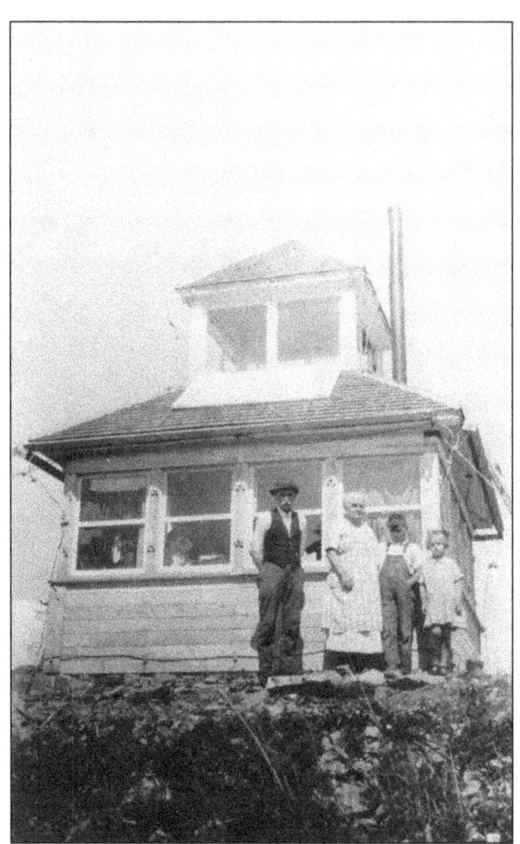

An unidentified family poses in front of the Tallowbox Mountain lookout in 1925. It was replaced with a 15-foot-tall tower in 1963. The tower was still in use by the Oregon Department of Forestry when vandals smashed windows and railings, then set a fire that damaged the support beams in December 2006. The lookout was torn down in 2007. (National Archives.)

A group of people pose in front of graffiti-covered rocks on Mount Ashland in this undated photograph. The lookout was built in 1922 and was used for 20 years before being abandoned. It was finally removed in 1959. Mount Ashland is now the site of a ski hill. (National Archives).

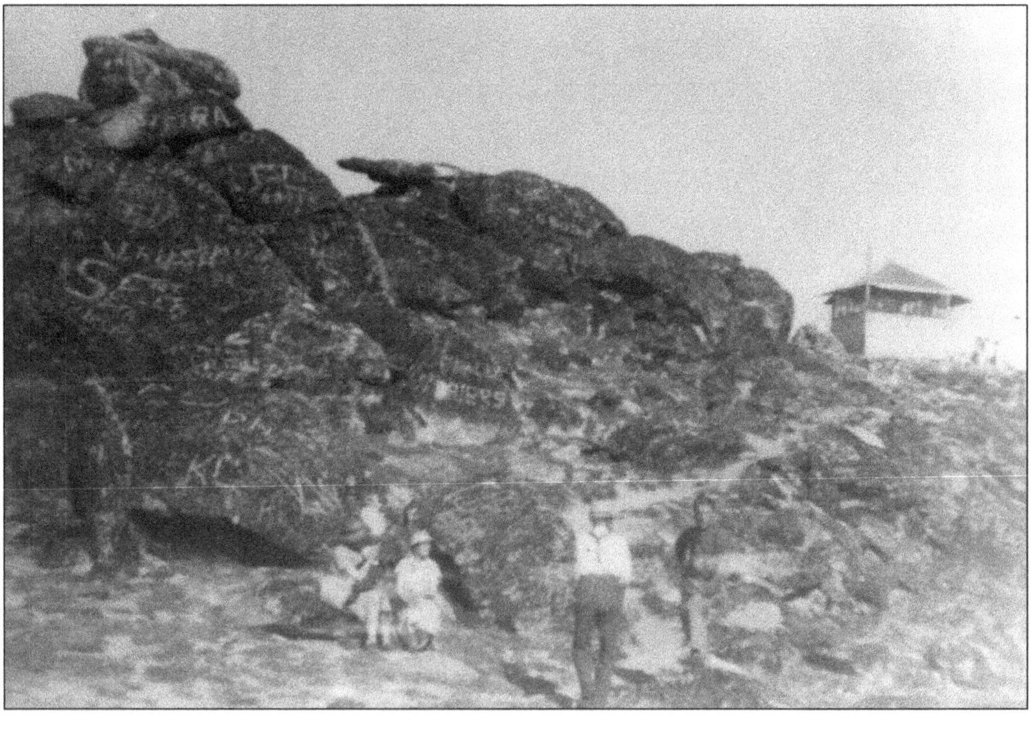

Early lookouts were often crude affairs, with a fire finder mounted on a stump or up a tree in a crow's nest and a phone affixed to a tree or pole. The observer stationed there would have to sleep in a tent nearby (or a cabin, if he was lucky). Pictured is Ernest W. Smith on Devil's Peak in 1915. (National Archives.)

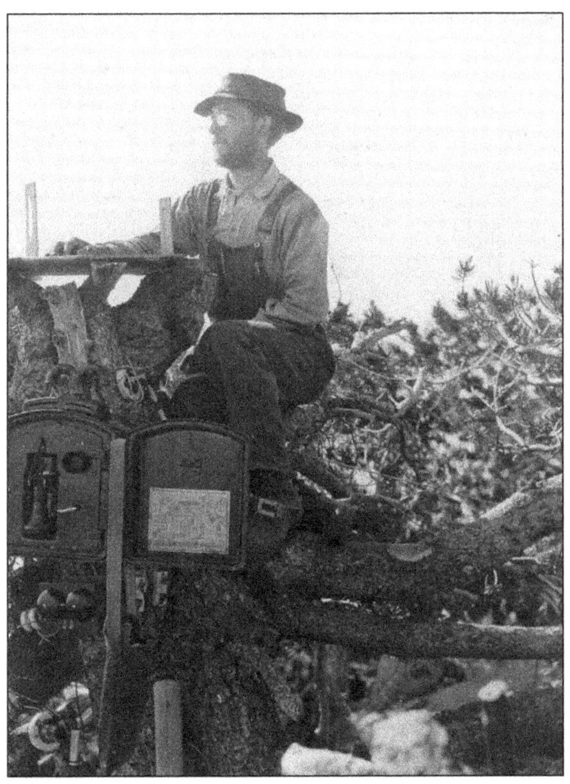

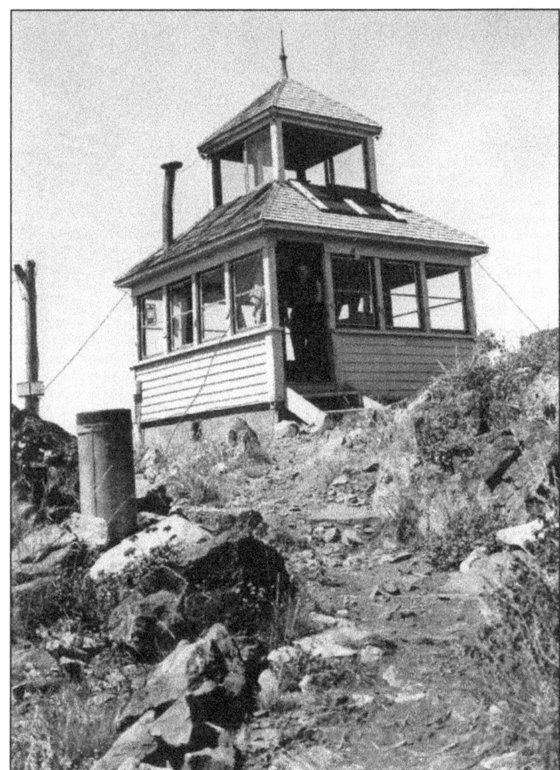

The cupola-style lookout on Dutchman Peak was built in 1927. Although this style of lookout was once very common, there are very few left standing. This one is still staffed every summer. The peak's name dates back to the 1870s, when a Dutch miner died in a snowstorm nearby. (National Archives.)

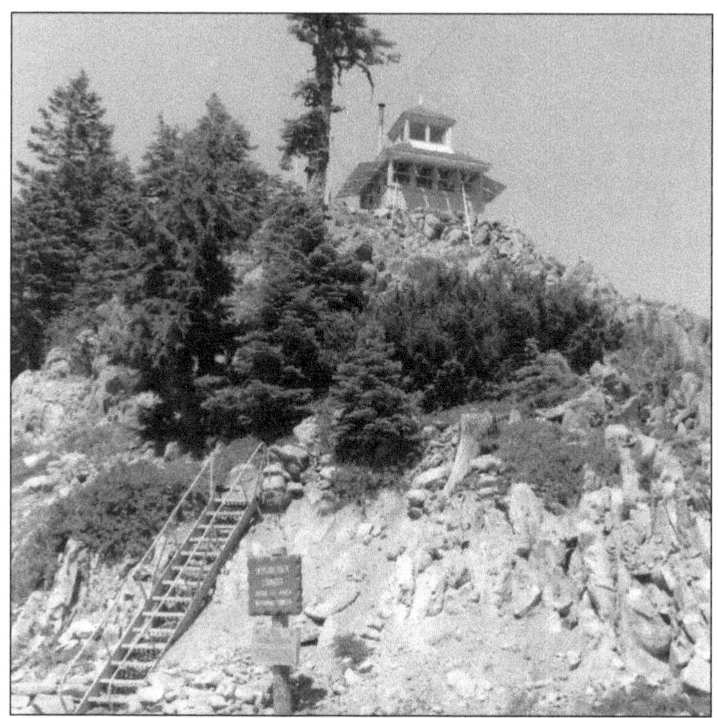

A D-6 cupola lookout was built on Hershberger Mountain in 1925. Although many other cupola lookouts were removed in Oregon over the years, this one survived. It fell into disrepair, but restoration work began in 2007. It is currently unused. (National Archives.)

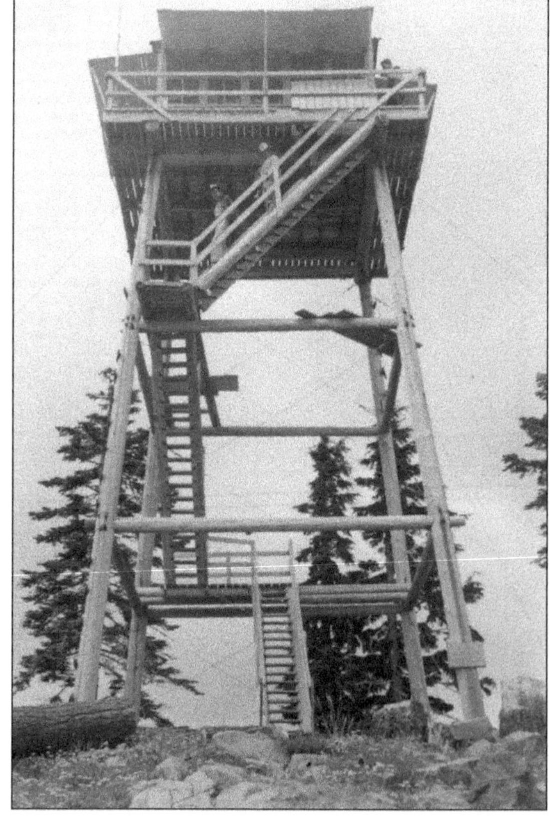

The first lookout on Blue Rock was this 45-foot-tall tower built in 1934. A new tower was built in 1963 but was moved to Robinson Butte 10 years later. The origin of the name "Blue Rock" is unknown but could derive from the basalt rock—which appears blue-gray in certain conditions—or from nearby Blue Lake.

Fire control officer Bob Webb and lookout Jim Weeks stand at the Osborne Fire Finder in the Cinnabar Peak lookout. The device was invented by US Forest Service employee William Osborne and has been in use at lookouts since 1915. With the fire finder, an observer can take an azimuth (directional bearing) on any smoke that is spotted and then report that bearing to dispatch. (National Archives.)

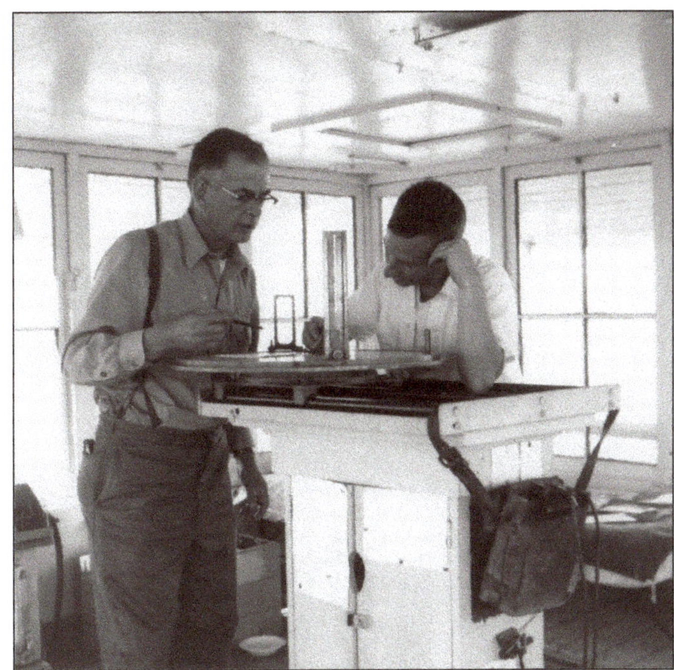

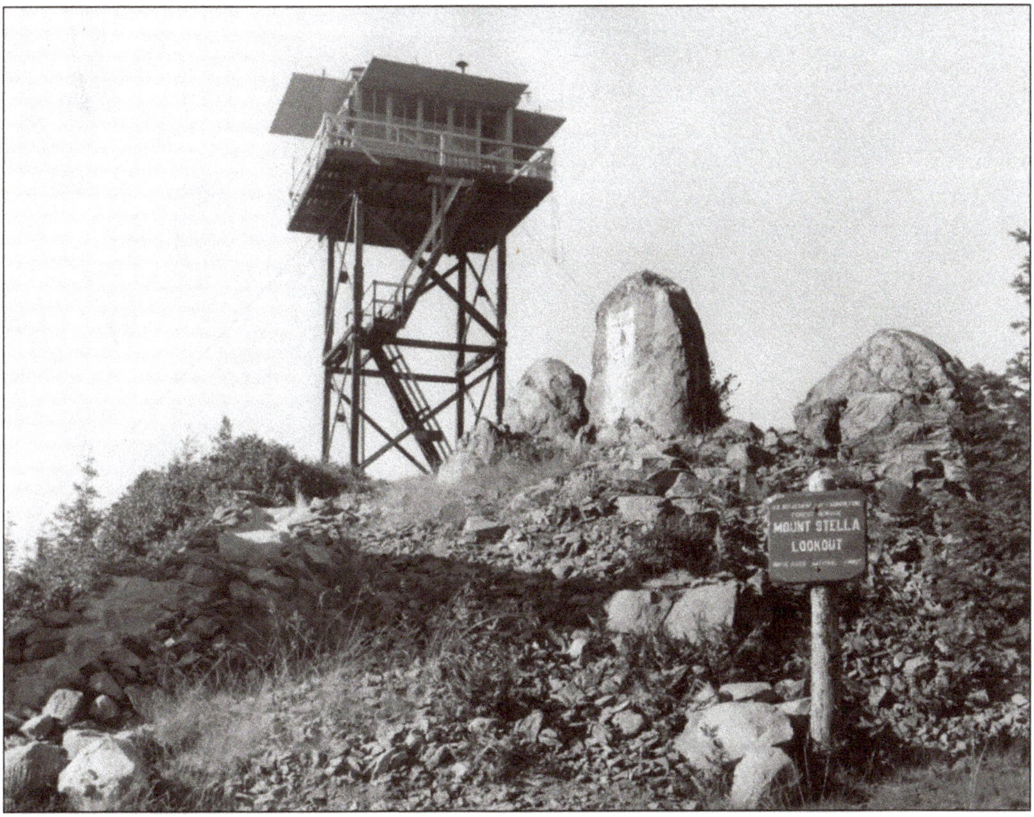

This 30-foot-tall tower on Mount Stella was built in 1946, replacing an earlier 1932 tower. It still stands, although it has not been used since the 1980s and is in bad shape. (National Archives.)

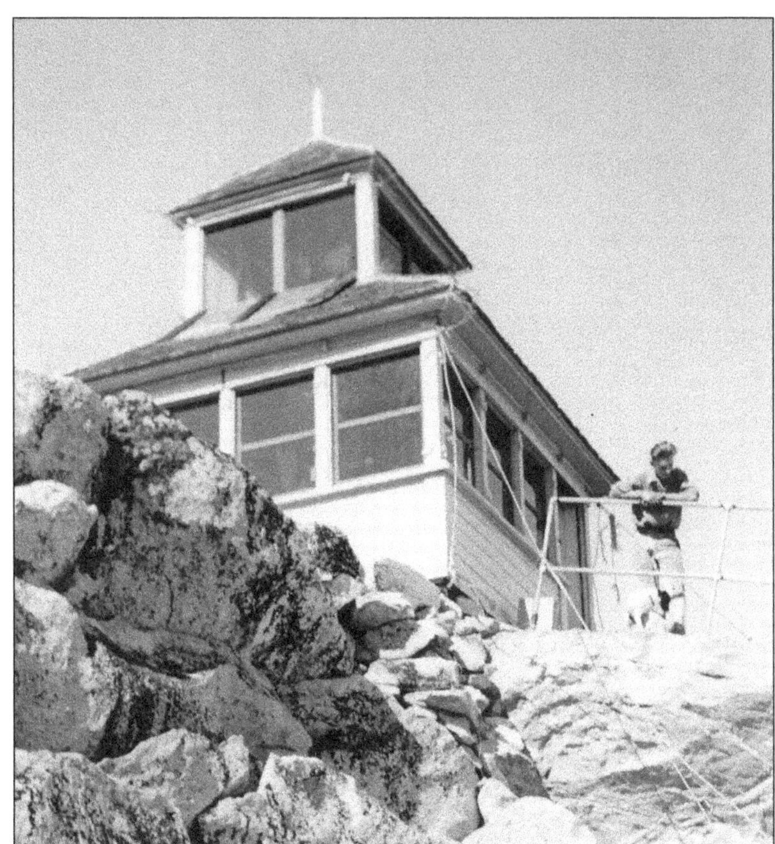

Merton LeRoy stands outside the Wagner Butte lookout, which was built in 1923. An R-6 cab replaced it in 1961 but was torn down 10 years later. (National Archives.)

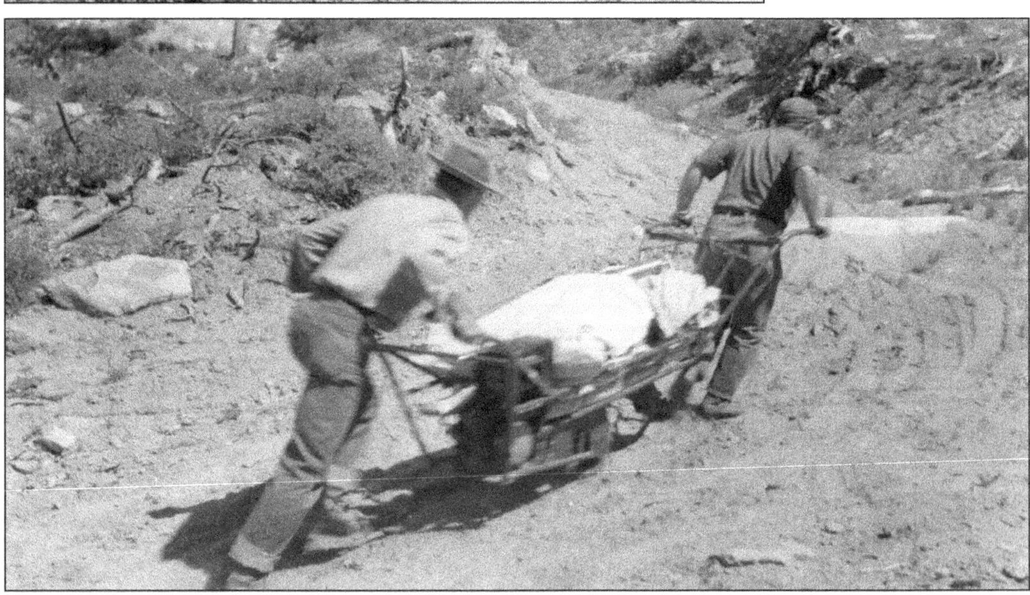

Two men push a gas-powered wheelbarrow on Wagner Butte. Such wheelbarrows had a 2.5-horsepower engine and could carry loads of 250 pounds. The Forest Service started using these machines during the late 1950s to ease strain on workers and to more easily move heavy materials on narrow trails.

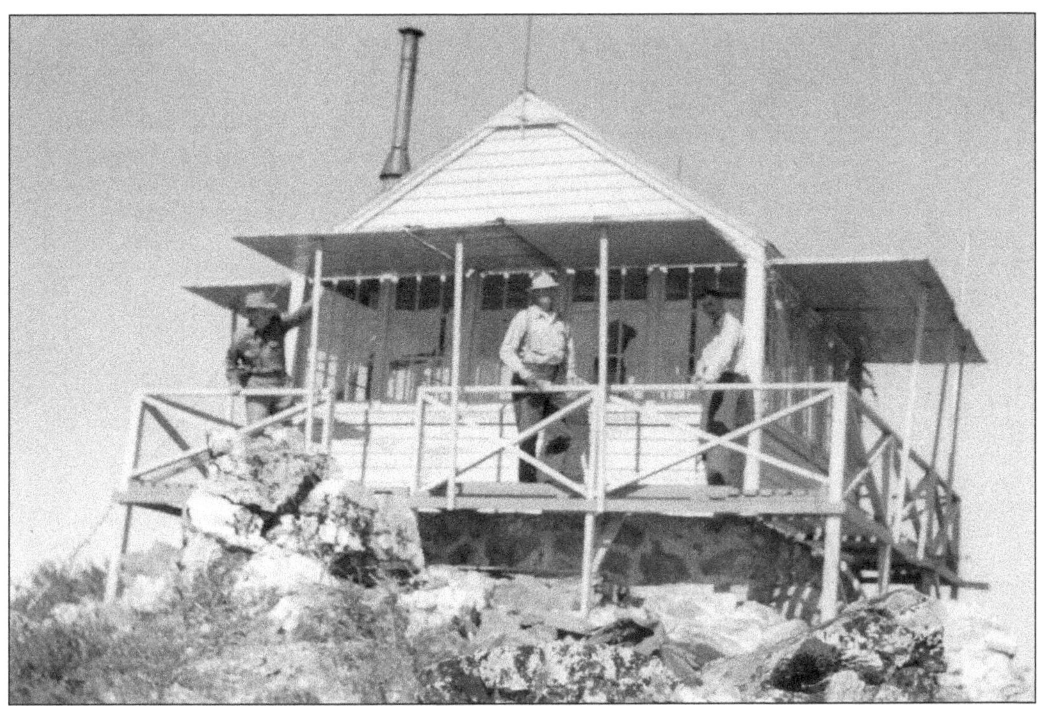

Pictured around 1945, this L-4 lookout was constructed on Whiskey Peak in 1930. During the winter of 1942–1943, Bill Zeigler was stationed here to watch for enemy aircraft as part of the Aircraft Warning Service. Twice during that winter, Zeigler's dog Two-Bits accidentally slid off the snowy cliff while chasing chipmunks. Both times, Zeigler thought the dog was a goner, and both times, the dog reappeared back at the lookout. (National Archives.)

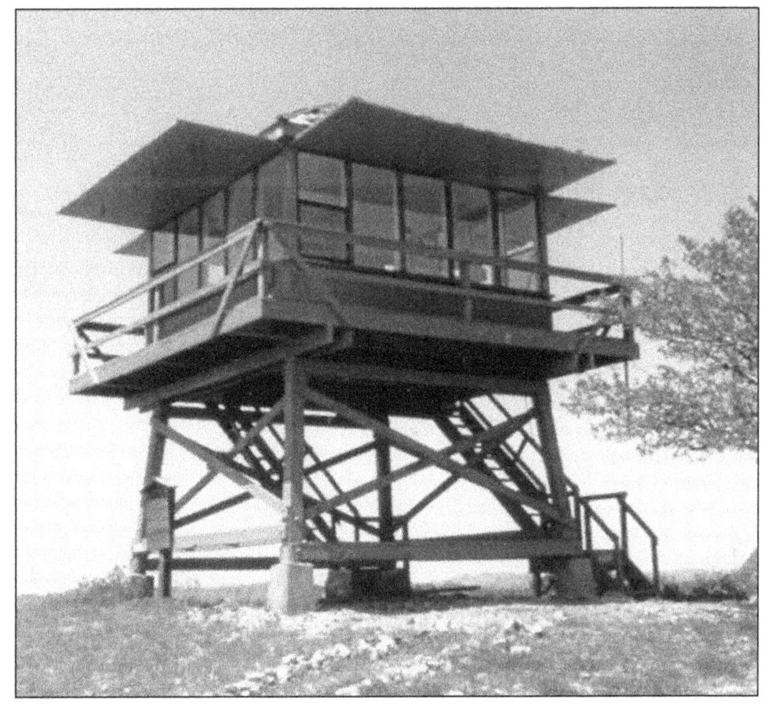

Anna Mae Winger staffed the lookout on Onion Mountain in the late 1960s and early 1970s. Because the lookout was so close to Grants Pass, her friends and family visited her frequently, and she never got bored. She also enjoyed the company of a ground squirrel named Harvey, which often came into the lookout to be fed by hand. (Ron Kemnow.)

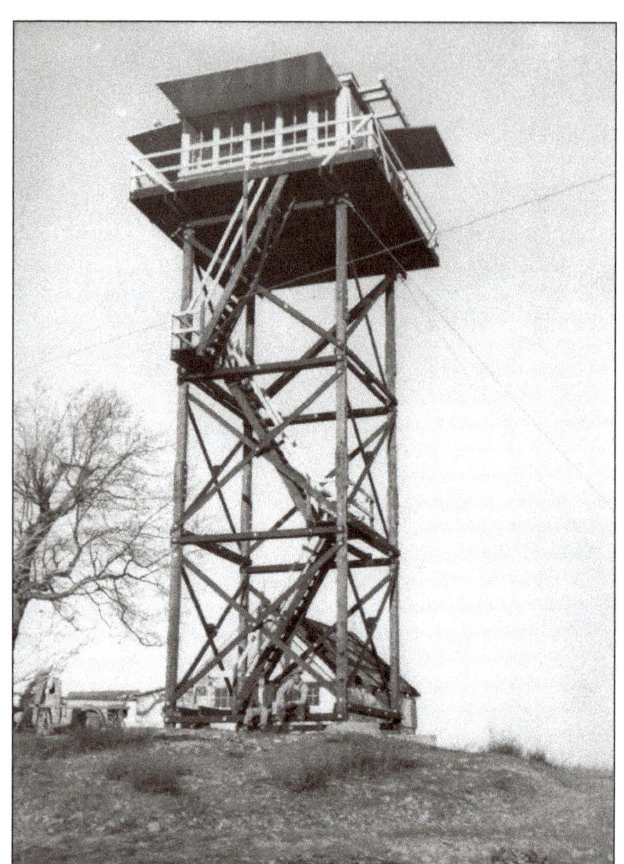

The Civilian Conservation Corps built a 50-foot-tall tower on Burnt Peak in 1933. In 1945, it was replaced by a 30-foot-tall tower with an L-4 cab (pictured) at a cost of $1,200. The lookout was removed in 1974. (National Archives.)

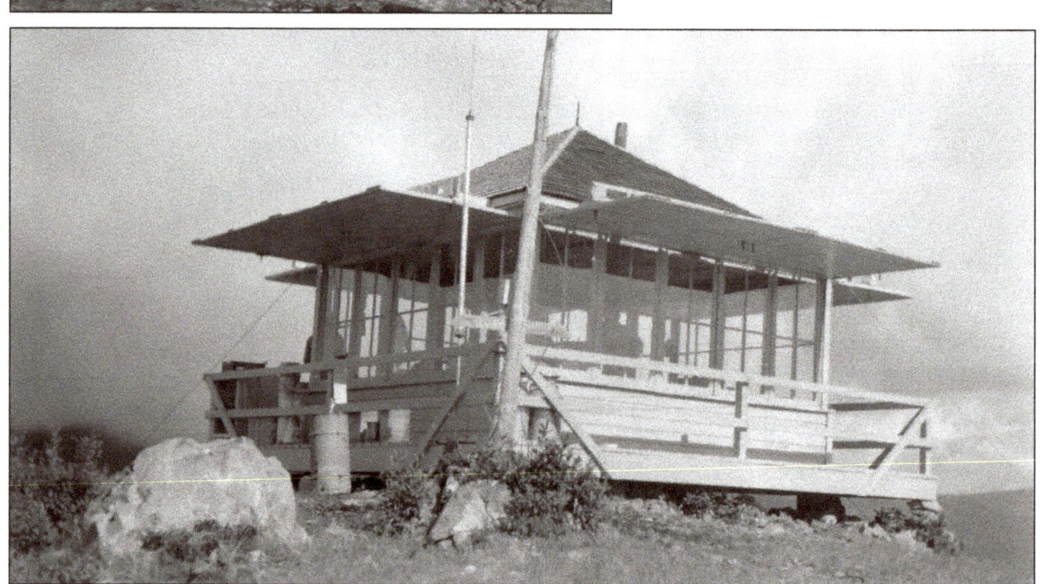

An L-4 ground cabin was built on Squaw Peak in 1943. It was built to provide improved coverage of the upper Applegate Valley area. It ceased to be used in the 1960s and has only served during emergencies since then. As of 2015, the Forest Service is finishing safety upgrades in preparation for making it available to rent. (National Archives.)

The Quail Prairie lookout was built in 1963 to replace the nearby Long Ridge lookout, which was destroyed in the Columbus Day Storm of 1962. It was available through the rental program but was damaged in a 2011 storm. The access road has also washed out, so the lookout has been abandoned. (Author's collection.)

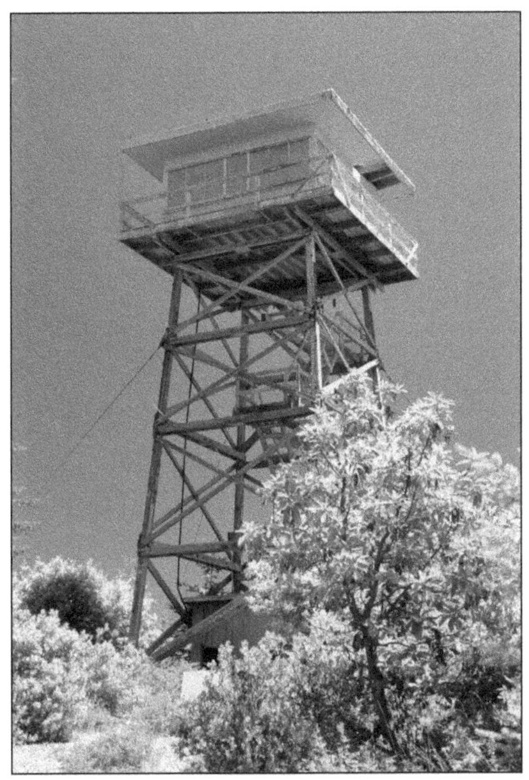

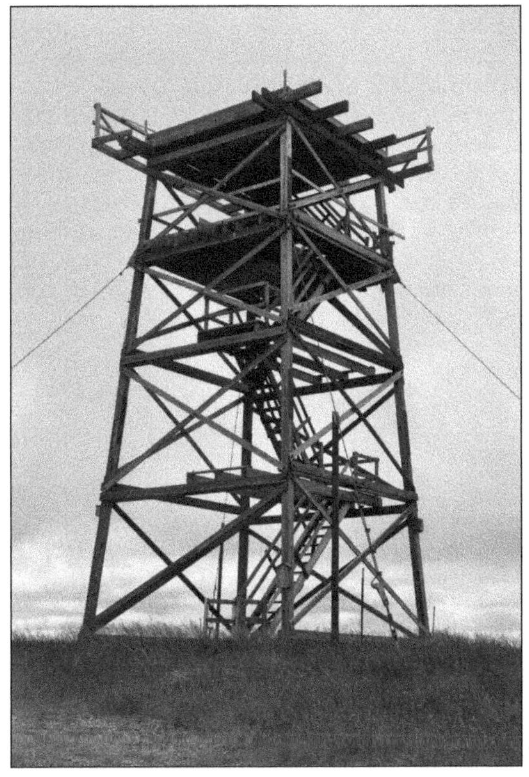

Wildhorse lookout was built in 1947, replacing an earlier 1935 tower. It was last staffed in the late 1990s before falling into disrepair. Heavy snow caused the cab to collapse during the winter of 2007–2008. As of 2015, repairs are underway to restore the lookout. (Author's collection.)

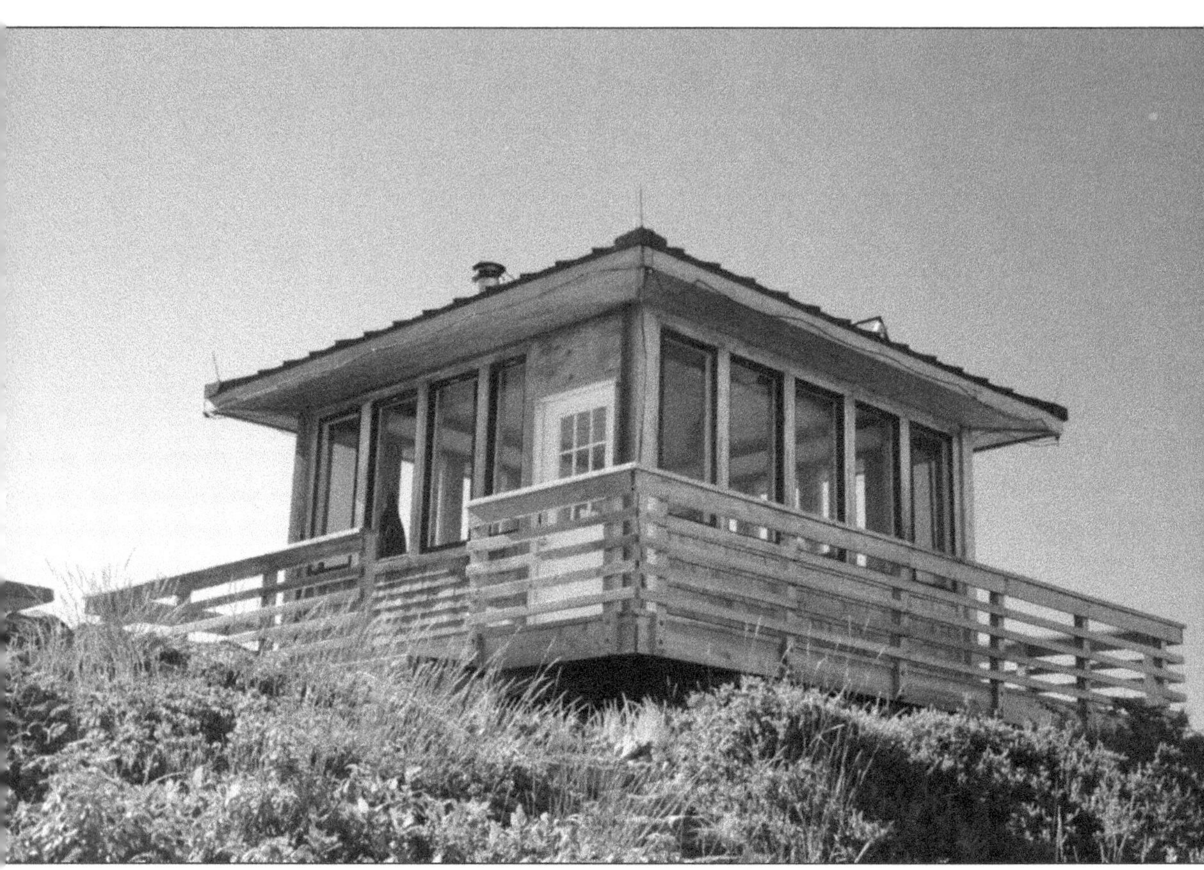

Snow Camp Mountain became a lookout site with an alidade and a tent camp around 1910. A 14-by-14-foot lookout cabin was built in 1924, replaced by an R-6 lookout in 1958. It ceased to be used during the 1970s but became the first Oregon lookout available through the rental program in 1990. It was extremely popular and often booked quickly. In July 2002, several lightning-sparked wildfires in southern Oregon joined into one massive fire that became known as the Biscuit Fire. On August 12, the lookout was consumed by the wildfire. The heat was so intense that the foundation cracked. In 2004, Don Hartley, a contractor from Crescent City, California, and others donated $40,000 worth of labor to help rebuild the lookout, which is pictured in 2014. (Author's collection.)

Six

Central Oregon

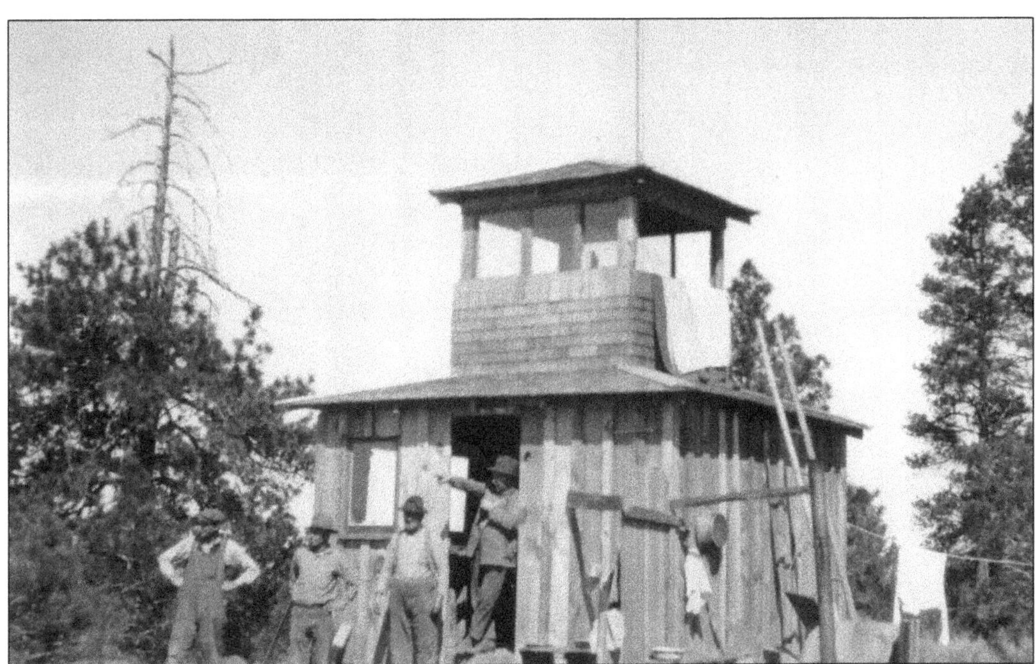

A 12-by-12-foot lookout cabin with a seven-by-seven-foot cupola was built on Emigrant Butte in 1921. It is pictured during the final stages of construction. A 20-foot-tall tower replaced it in 1933 but was removed in the 1940s.

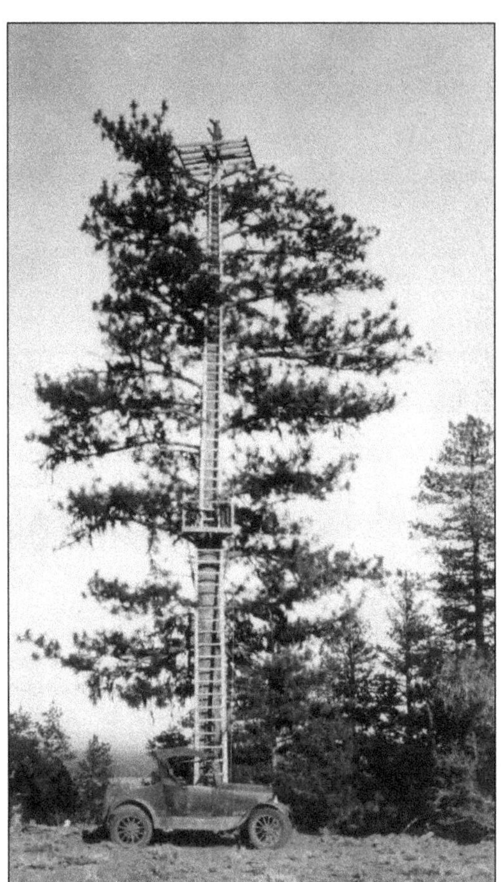

A crow's nest was built 70 feet up a tree on Black Mountain in 1937, along with a ground cabin. The cabin has since been moved to the Ochoco Ranger Station, but remnants of the ladder and crow's nest still remain in the tree.

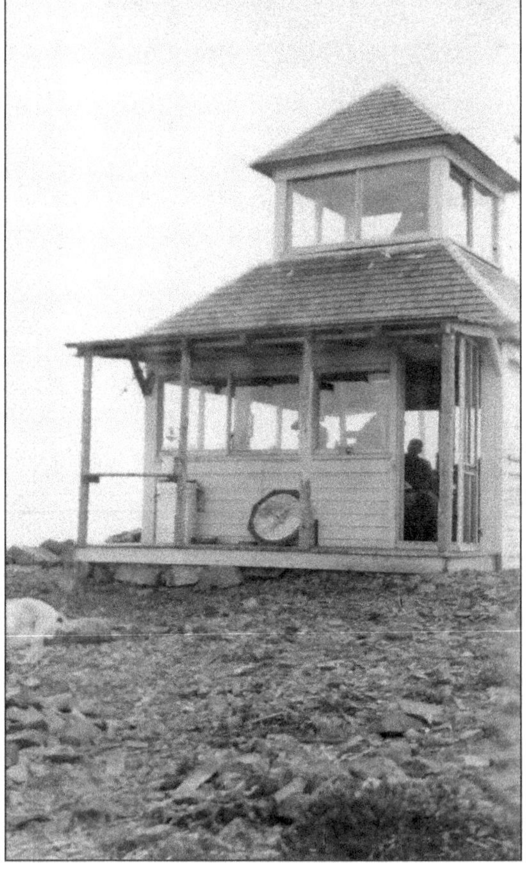

Jack Groom worked at the Dixie Butte lookout in the early 1930s. His first summer, he hiked seven miles to the lookout, and a few days after arriving, the ranger called him up on the radio and said there was a problem because Groom did not turn 18 until the end of July. The ranger decided to take a chance and let Groom stay, much to his relief, since the work was much appreciated during those Depression years.

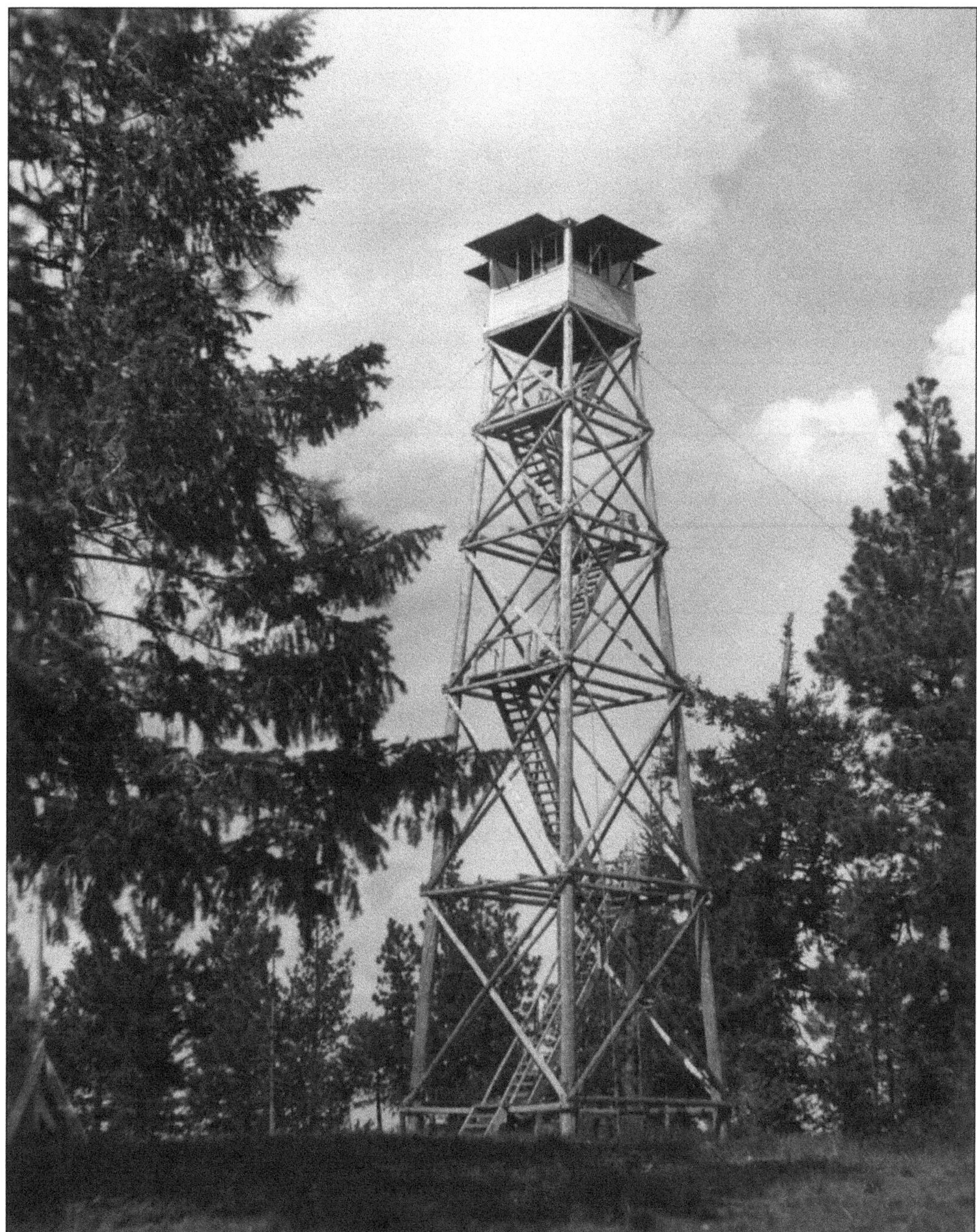

Lookouts were sometimes treated to nighttime astronomical events. In the early morning hours of August 8, 1971, a meteor entering the earth's atmosphere above Oregon created a fireball visible to people throughout the state, including Denise Linae Myers, who was stationed at the Foley Butte lookout. She was quoted in the *Oregonian* as saying, "The tail was a golden bunch of sparkles—really beautiful." This photograph shows the first Foley Butte lookout, which was built in 1932. It was replaced with a new tower in 1953, which still stands.

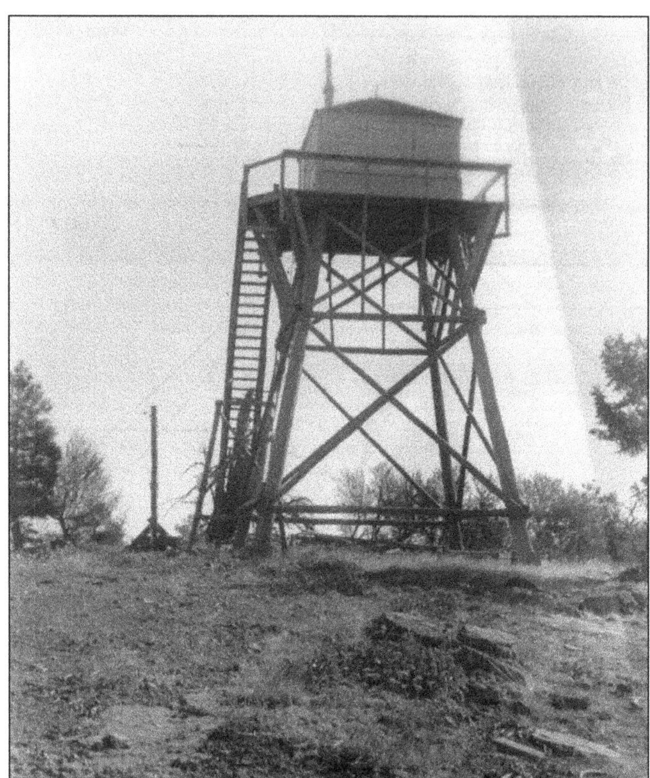

A 25-foot-tall lookout was built on Hash Rock in 1935, replacing a crow's nest lookout built three years earlier. The tower was removed in the 1960s. The summit was named after Allan Hasch, an early settler in central Oregon who may have had a mining claim nearby.

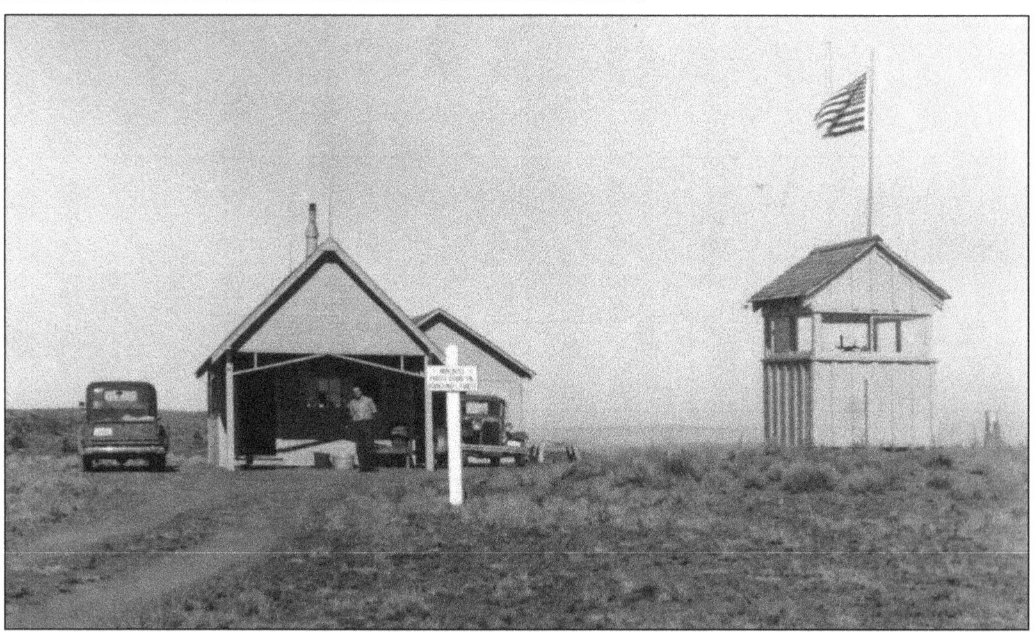

Mackey Butte became a lookout site in 1925. In 1935, a cabin and garage were built. The fire finder sat out in the open until a small six-by-six-foot cabin was built around it in 1940. The site was known as Norcross Guard Station, named after nearby Norcross Creek, which was in turn named after early settlers Jacob and Ellie Norcross. Today the site is on private property, but the buildings still stand.

Linda Shaw worked at the Gerow Butte lookout during the summers of 1962 and 1963 as a way to earn money for her college education. She is pictured standing in front of the ground cabin that served as her living quarters. (Linda Shaw.)

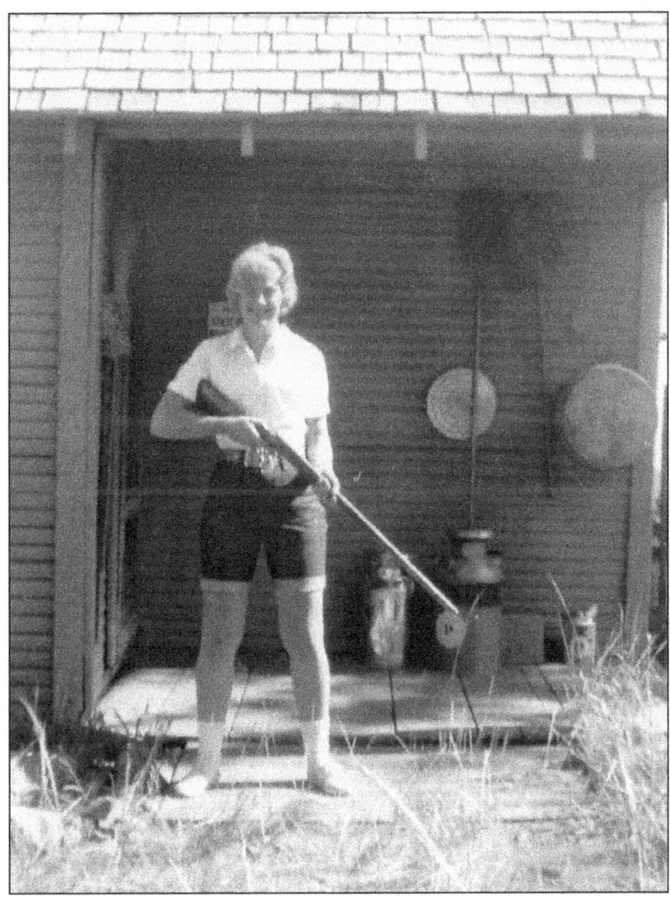

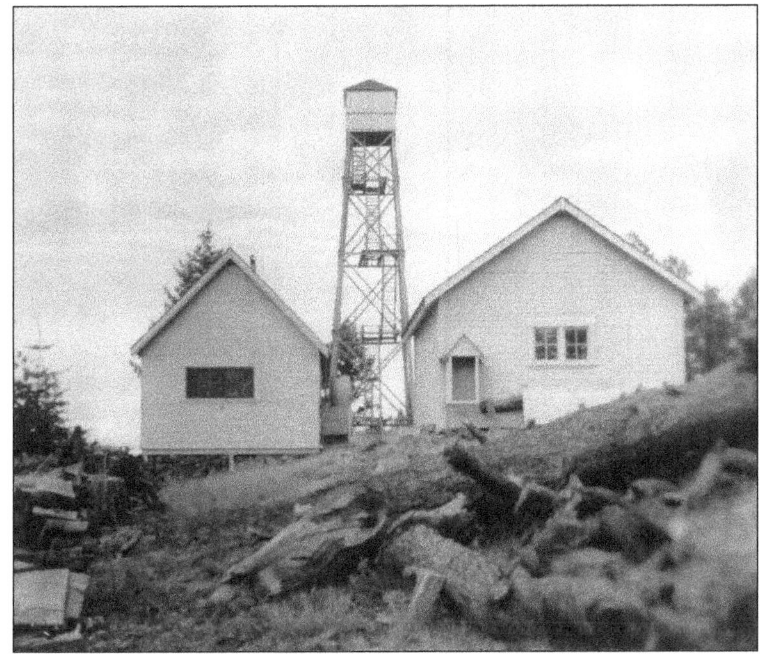

Although the Gerow Butte lookout was never struck by lightning during Linda Shaw's two summers there, it was almost taken out by a fighter jet on a practice run. The jet was flying low and evidently did not know about the 60-foot-tall tower. As the plane flew by—just 50 feet away —she could see the pilot's face in the cockpit.

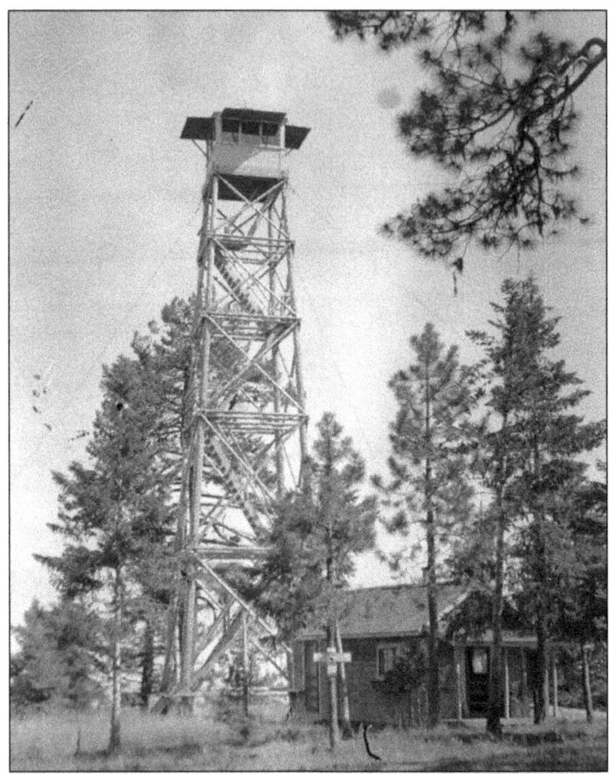

A 1926 crow's nest on Ingram Point was replaced by the 50-foot-tall wooden tower (pictured) in 1934. The tower was removed in the 1960s. The mountain is named in honor of Douglas C. Ingram, a Forest Service employee who was the first lookout to be stationed here. He died in a wildfire in Washington's Chelan National Forest in 1929.

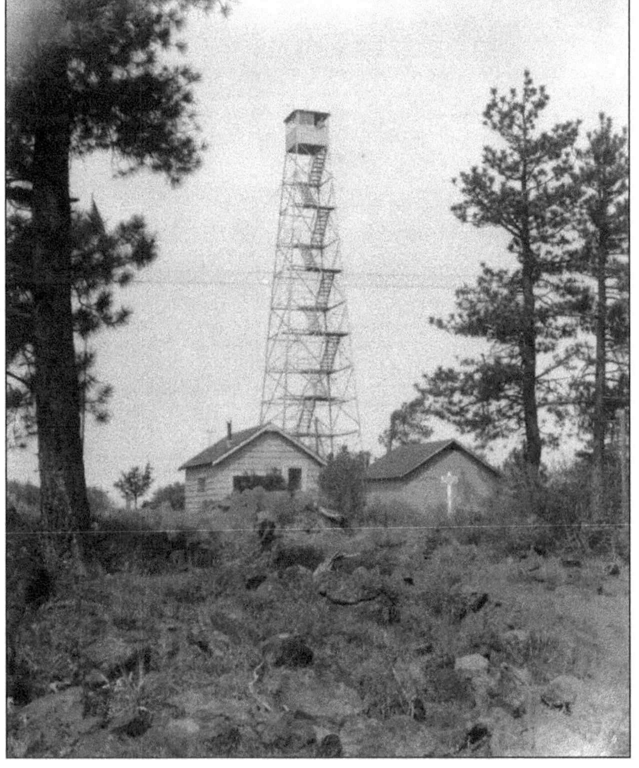

While working as a lookout on Dry Mountain in the 1950s, Benny Parmele would see fighter jets flying by on practice runs. They flew so close that they would salute him as they flew by, and he would salute them in return. Another lookout remembered there being a lot of rattlesnakes at Dry Mountain. When Ranger Cal Weisenfluh came to visit once, the lookout's puppy snuck up on him and tugged at his pants leg, scaring the ranger, who thought it was a rattler.

Two unidentified men pose in front of the Bald Butte lookout in 1940. The 45-foot-tall tower and ground cabin was built in 1931 and replaced in 1959. That tower still stands but has been abandoned.

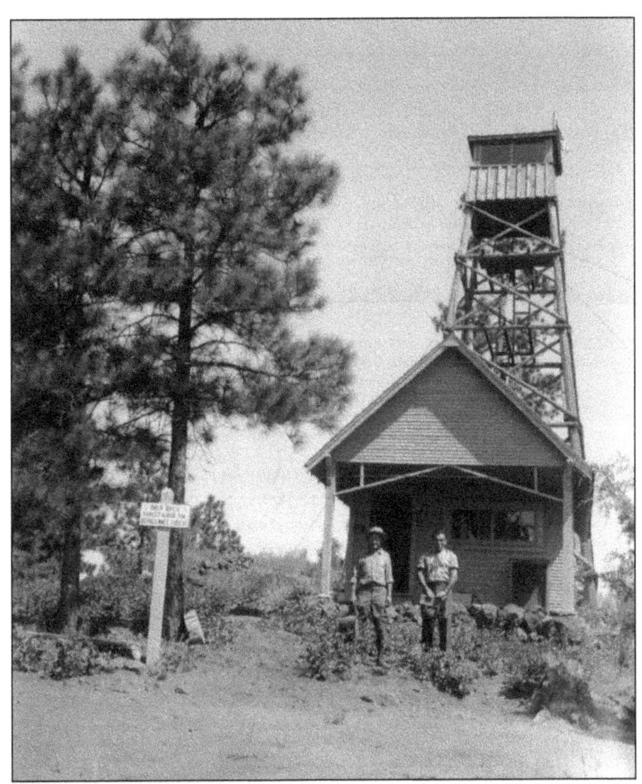

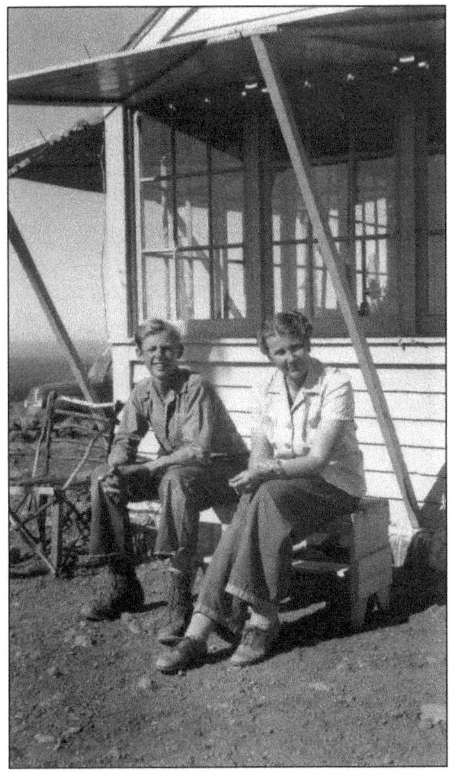

During World War II, many lookouts were staffed by women, because the young able-bodied men who had traditionally served as lookouts were away at war. This 1944 photograph shows Seattle schoolteacher Marjory Grant and her nephew Clive Ingalls at the Snow Mountain lookout. They reported 12 fires from this lookout that summer.

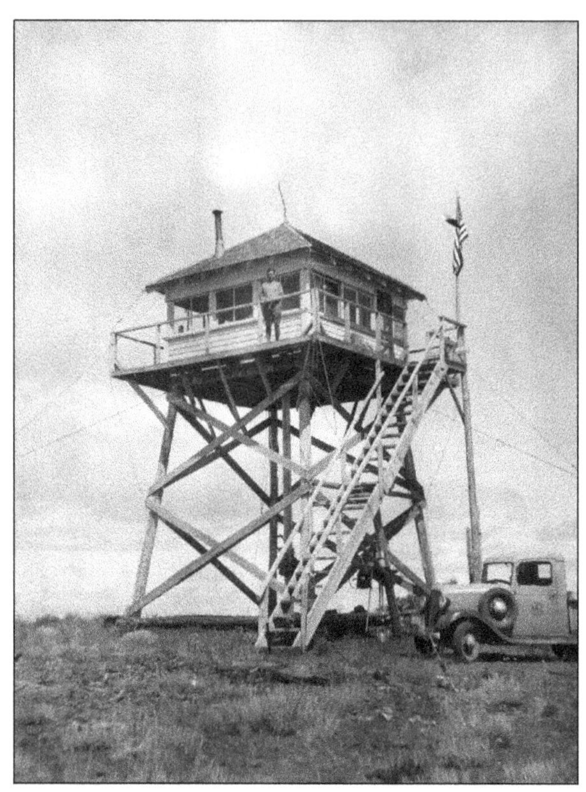

A 14-by-14-foot cabin was built on Spanish Peak in 1921 and then raised onto a 20-foot-tall tower in 1931, pictured on the left. Mabel Pengelly, pictured below, was a schoolteacher from Pendleton who spent the summer of 1944 working the Spanish Peak lookout. She is pictured standing next to the Osborne Fire Finder, which was a fixture in every lookout.

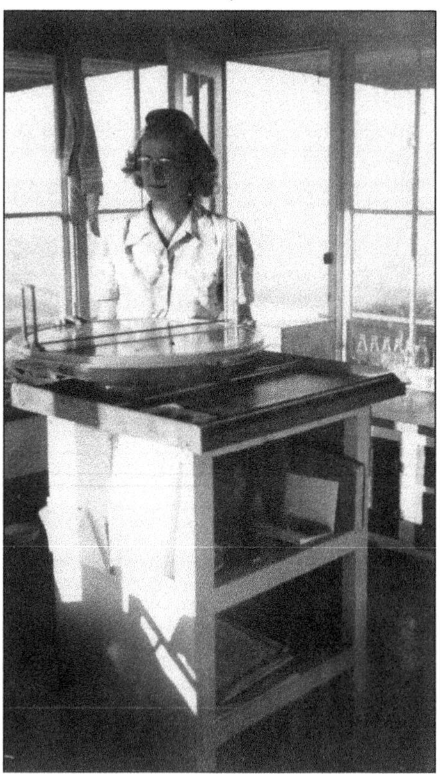

The 50-foot-tall tower pictured was built on Wolf Mountain in the 1920s. A new 107-foot-tall tower was constructed in 1947. It still stands today and is staffed every summer. It is the tallest standing tower in Oregon.

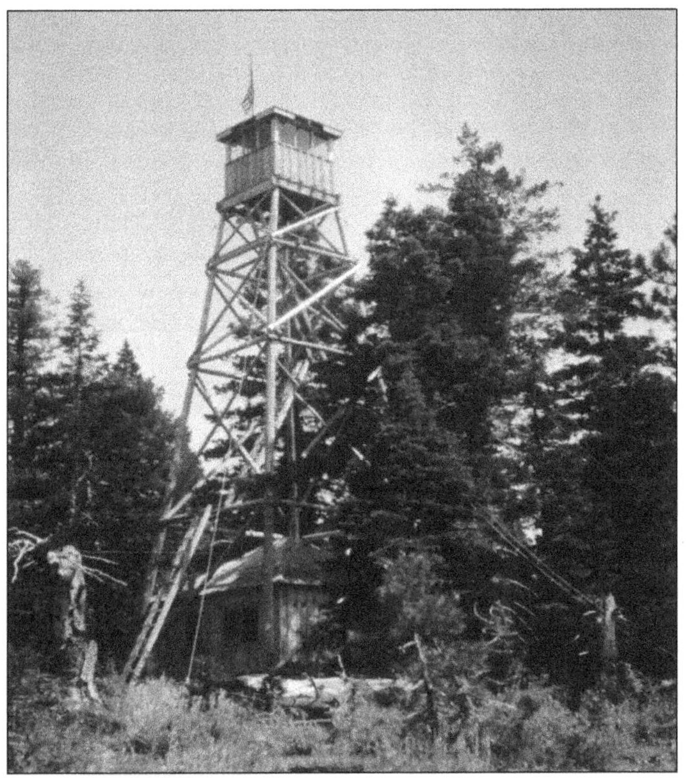

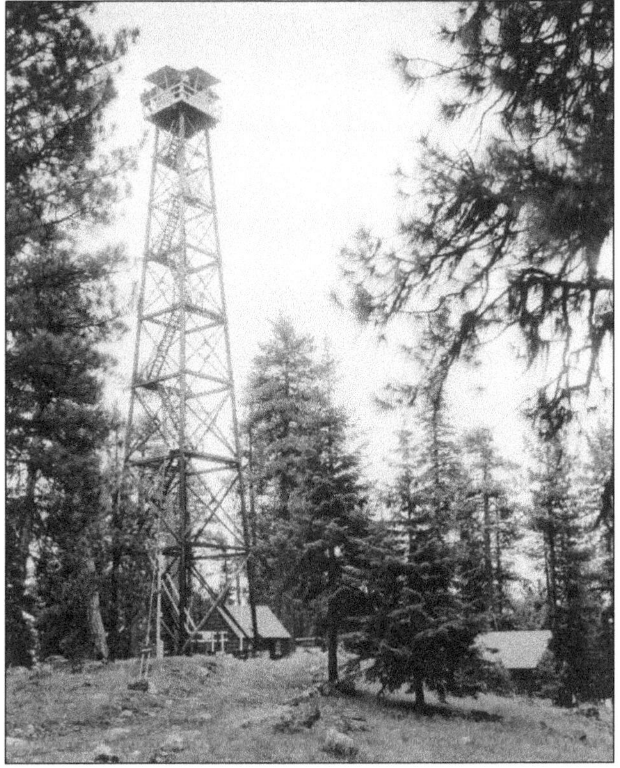

The 100-foot-tall lookout on Frazier Point was built in 1936. Working in a tower that tall was not for everyone. One young man shouted into the radio that he was going to "fly away now," and when a helicopter crew was dispatched to check on him, he was found sitting at the bottom of the stairs with a blank look, mumbling to himself. The cabin burned down in the 1970s, but the tower still stands, although it is no longer staffed. (National Archives.)

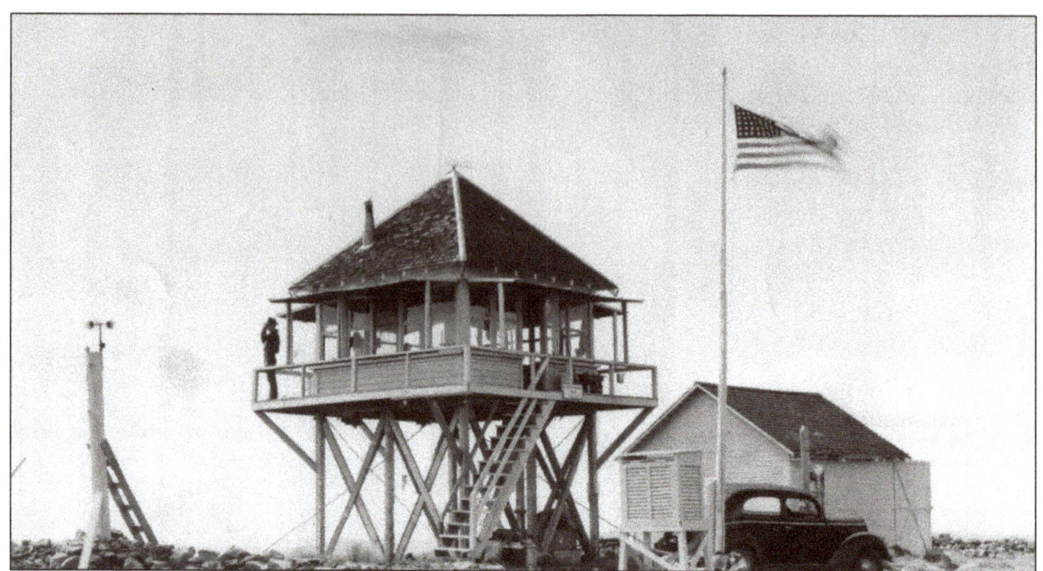

Ranger Alex Donnelly built the first lookout cabin on Crook County's Lookout Mountain in 1912. It was replaced by a 10-foot-tall tower in 1925 (pictured). In 1928, Gordon McNely asked for permission to leave his station on Lookout Mountain because he felt the "flu" coming on. He returned pretty quickly and did not return alone, accompanied by his brand new bride.

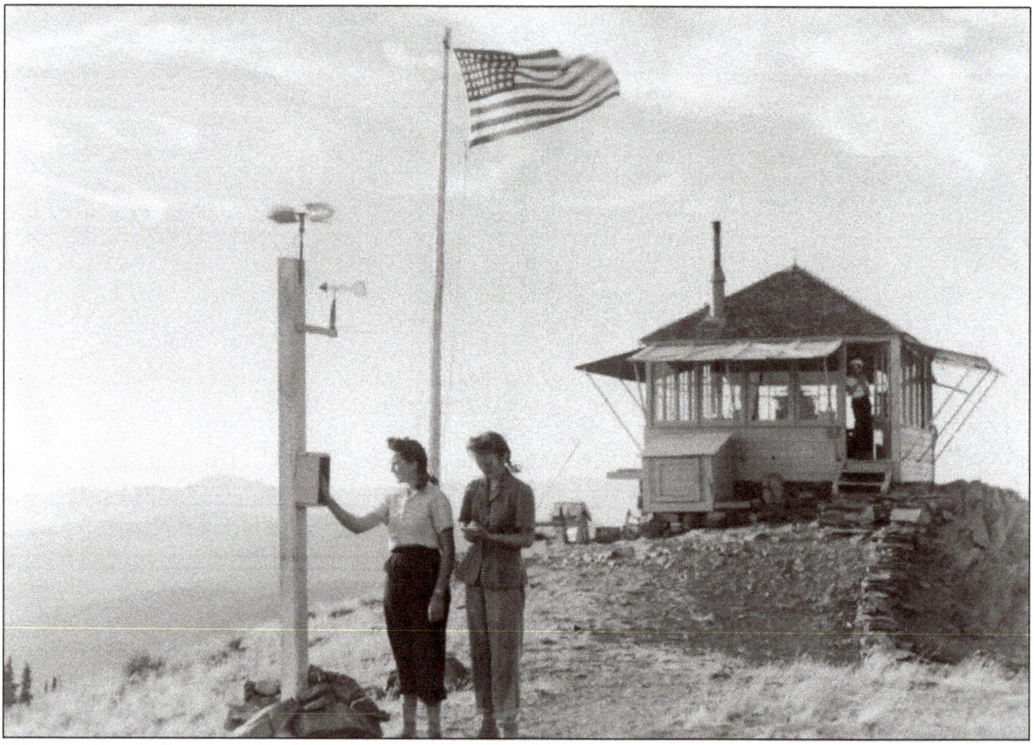

This undated photograph shows two women checking a weather reading on Baldy Mountain in the Malheur National Forest. One of the duties of a lookout was to make daily reports on wind speed, precipitation, relative humidity, cloud cover, and more weather-related measurements. The 1933 L-4 lookout on Baldy Mountain was removed in 1968.

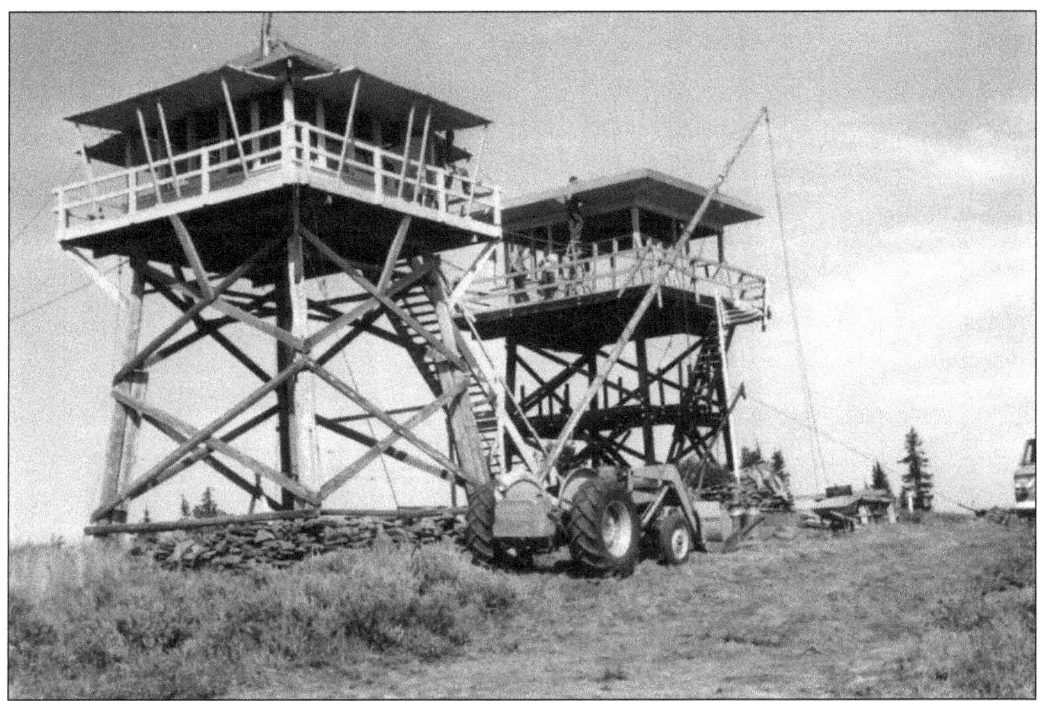

Mount Pisgah has been a lookout site since 1918. Pictured is the 1933 L-4 tower (left) and the new R-6 tower (right) under construction in 1963. This tower still stands and is staffed every summer.

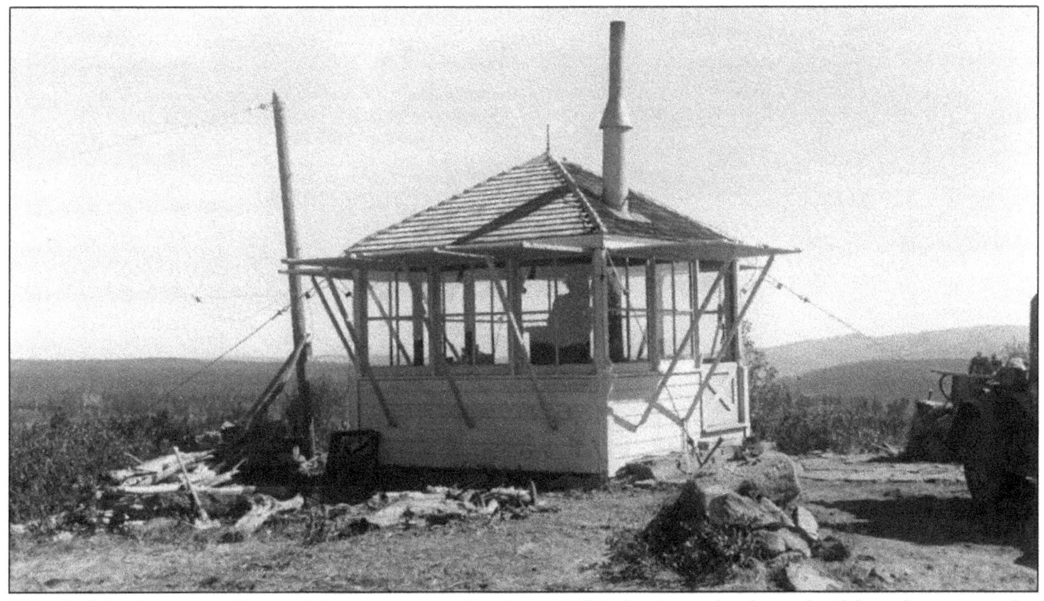

The L-5 lookout on Burnt Mountain was built in 1932 during a lookout building boom in the Malheur National Forest. It was removed in 1968.

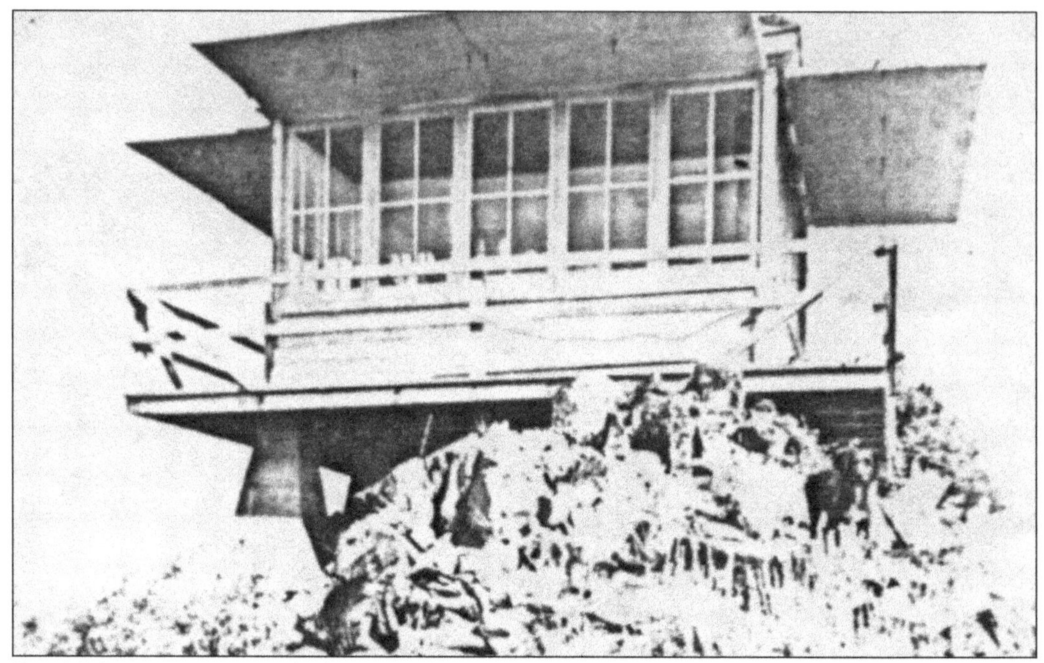

Don Nolen was serving as lookout at West Maury in 1958 when a bear tried to get into the cabin in the middle of the night. Nolen didn't have his glasses on but got out his rifle and shot at the bear through the open window. The bear left but came back, and Nolen shot at him again. In the morning, the bear was found dead under a nearby tree.

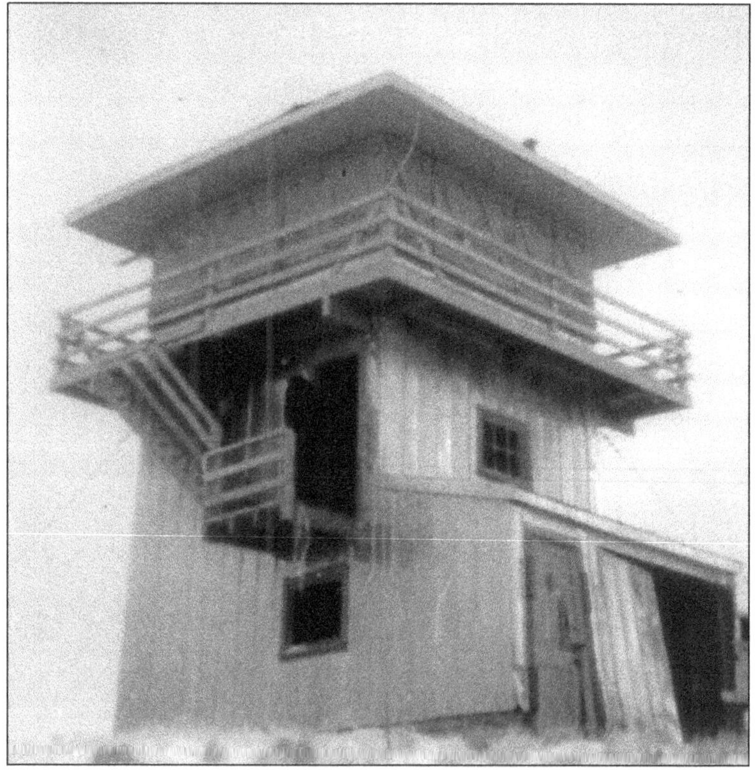

Ray and Ella Moles worked at Pyramid Point in 1961. The three-story lookout had an observation room on top, sleeping quarters below, and storage on the ground floor. They had a few amenities that made lookout life easier, including a shower constructed from boards with a barrel on top, a gas-powered washing machine, and a gas-powered refrigerator. (ODF.)

Seven

EASTERN OREGON

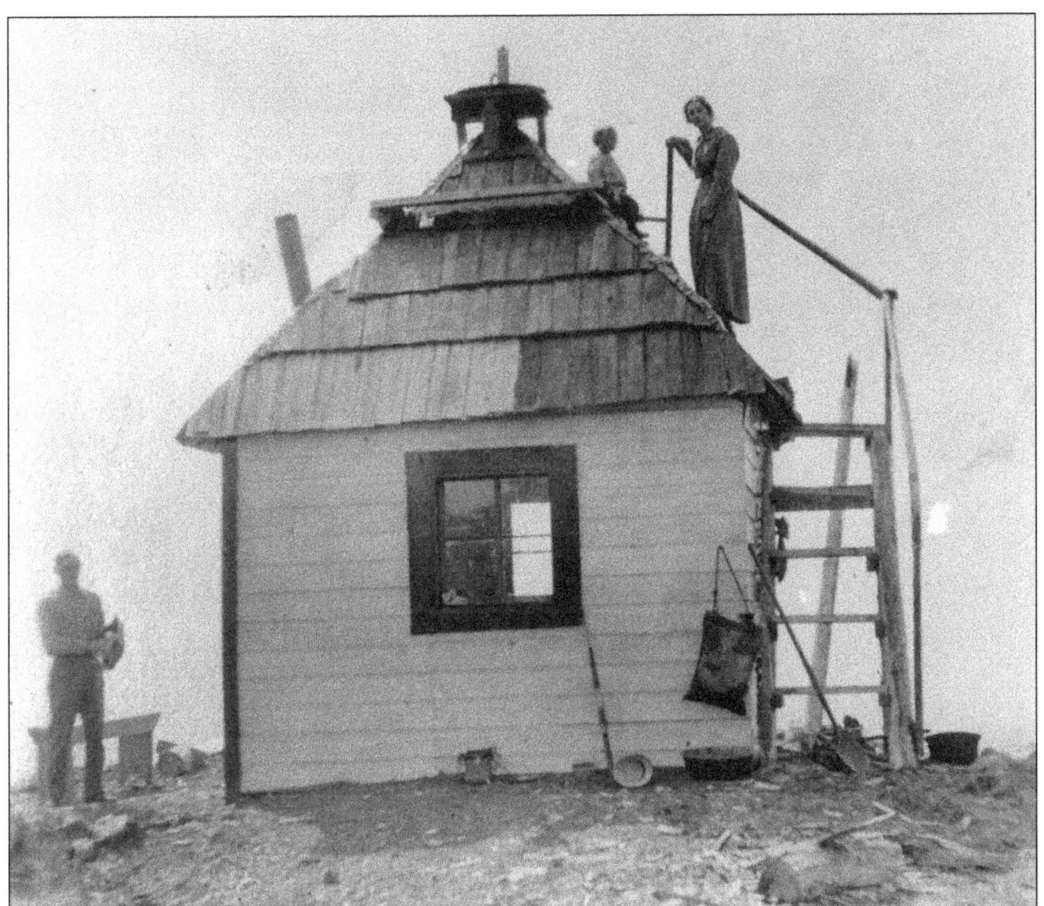

A cabin was built on Bennett Peak in the 1920s. It had a fire finder on the roof, reachable by a set of stairs on the outside of the cabin. The cabin was later replaced by a tower, but Bennett Peak ceased to be used as a lookout site by 1948.

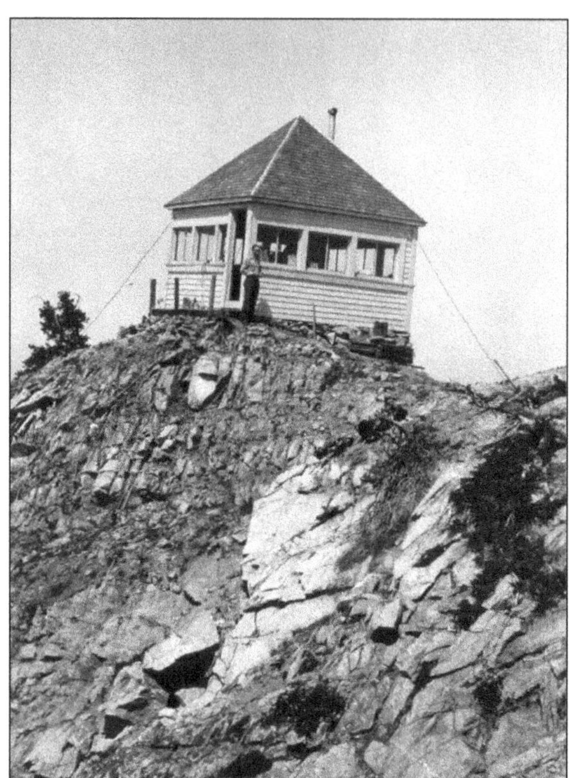

This D-5 lookout was built on 8,650-foot Cornucopia Peak in the 1920s. The lookout is long gone, and the peak is now within the boundaries of the Eagle Cap Wilderness, visited only by those willing to undertake an off-trail scramble.

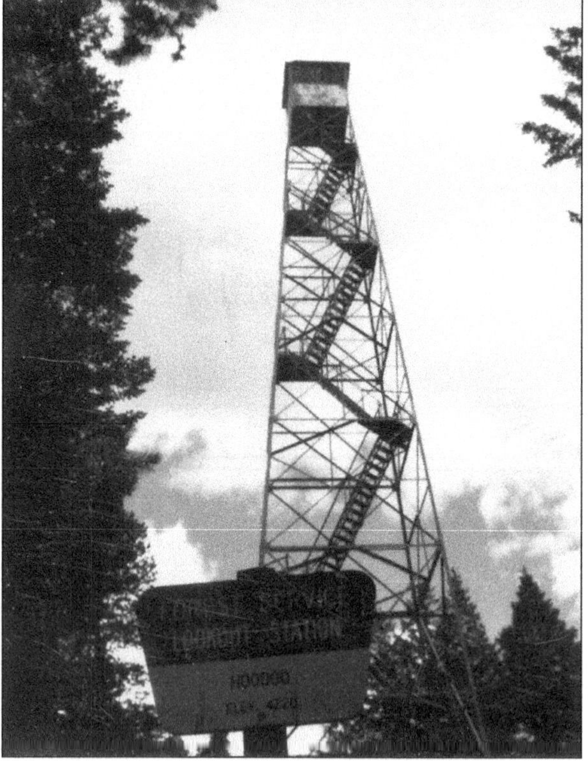

The 101-foot-tall steel Aermotor tower on Hoodoo Ridge was built by the Civilian Conservation Corps in 1933. It still stands, although it is no longer in use. Along with accompanying buildings at the site, it was listed in the National Register of Historic Places in 2015. It is the one of the tallest standing towers in Oregon.

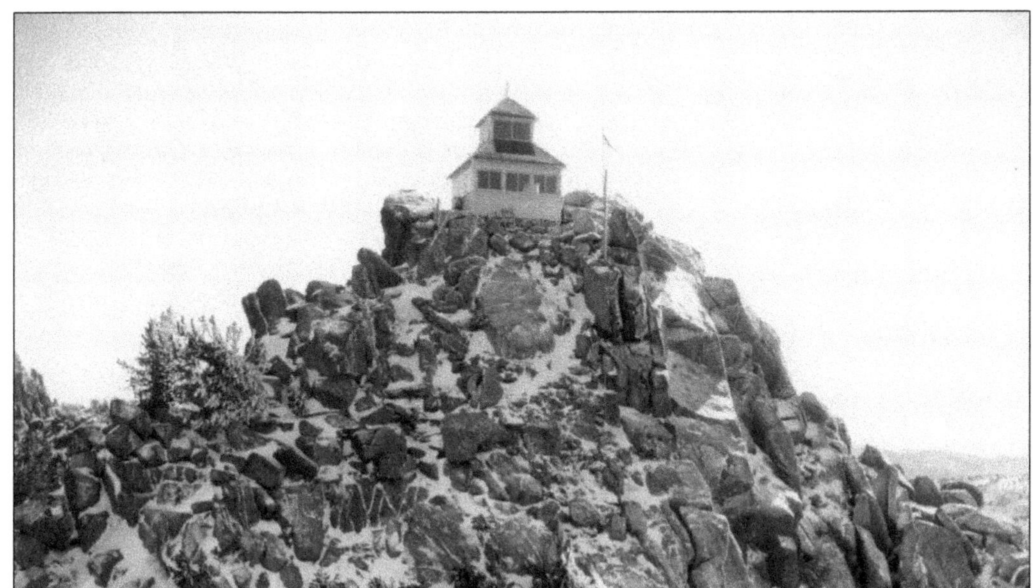

In 1912, a cabin with a rooftop platform was built on a mountain overlooking Anthony Lake. It was called the Lakes Lookout, but a cupola cabin (pictured) replaced it in 1926. Paul Ingram (pictured on the right in 1954) was stationed there in 1953 and 1954 and brought along his springer spaniel, Midge. Because he had to carry his supplies with him on his way to the job site, Ingram learned to make do with less, rationing his chocolate bars and making five gallons of water last him a full week (since the nearest water supply was a quarter mile away). The lookout was removed in the 1960s, although the spot remains a popular hiking destination. (Above, USFS; right, Paul Ingram.)

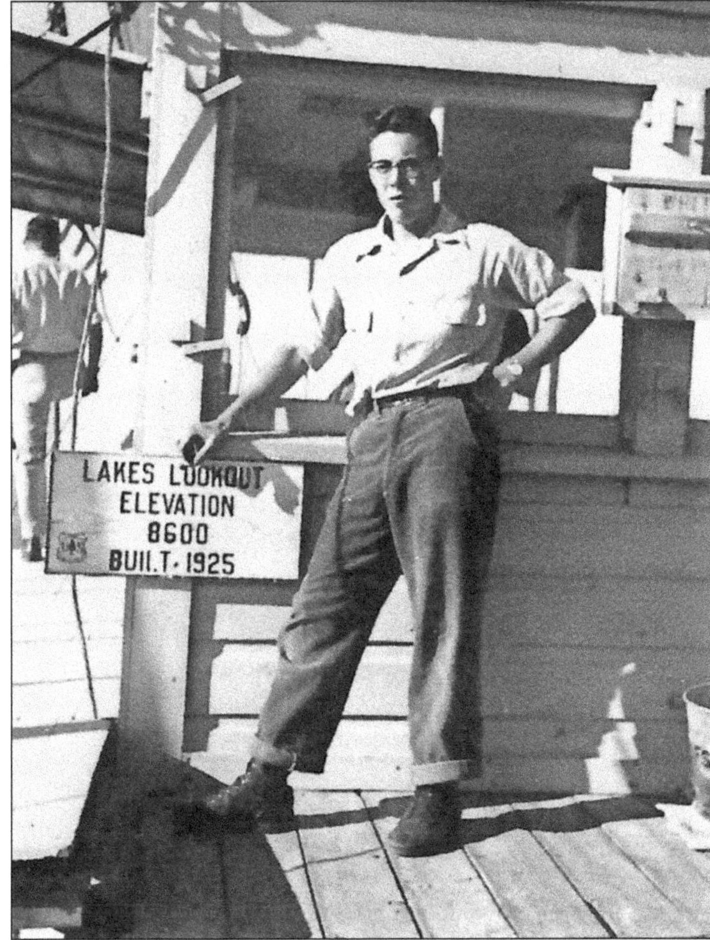

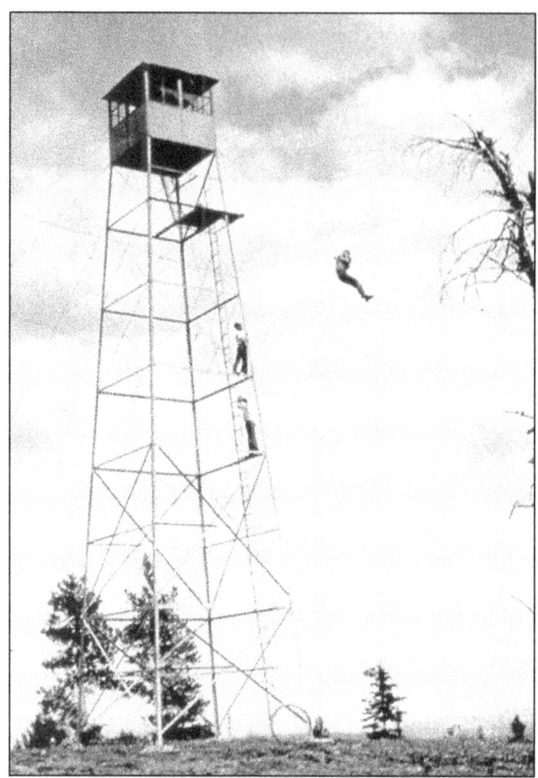

This 50-foot-tall steel tower was built on Desolation Butte in 1923. District Ranger Lloyd Waid recalled seeing a group of young people who played a risky game on the tower. They would climb to the top with a gunny sack, which they would then use to wrap around a guy-wire and ride it down to the ground, as pictured.

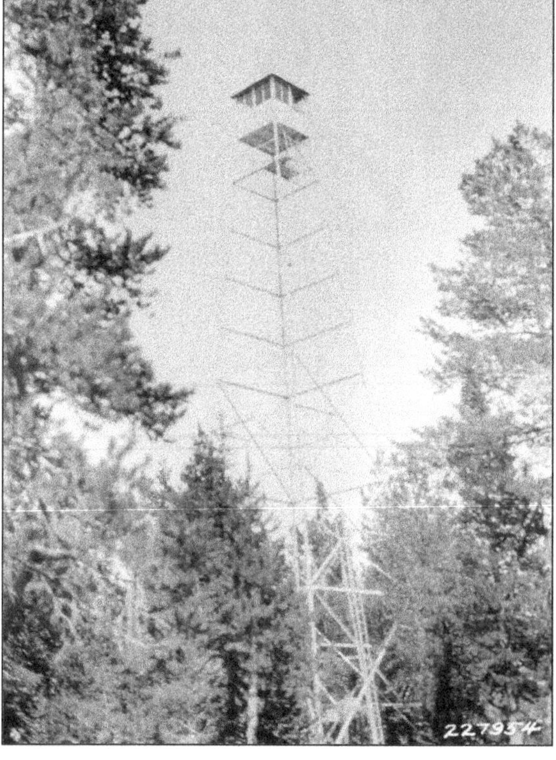

Lloyd Woodell was stationed at the Mount Emily lookout in 1927 when he decided to hunt some squirrels. Before he could shoot anything, though, he was shot himself by a careless man firing from the woods. The man ran off, and Woodell was fortunate that his arm was only slightly injured by the bullet.

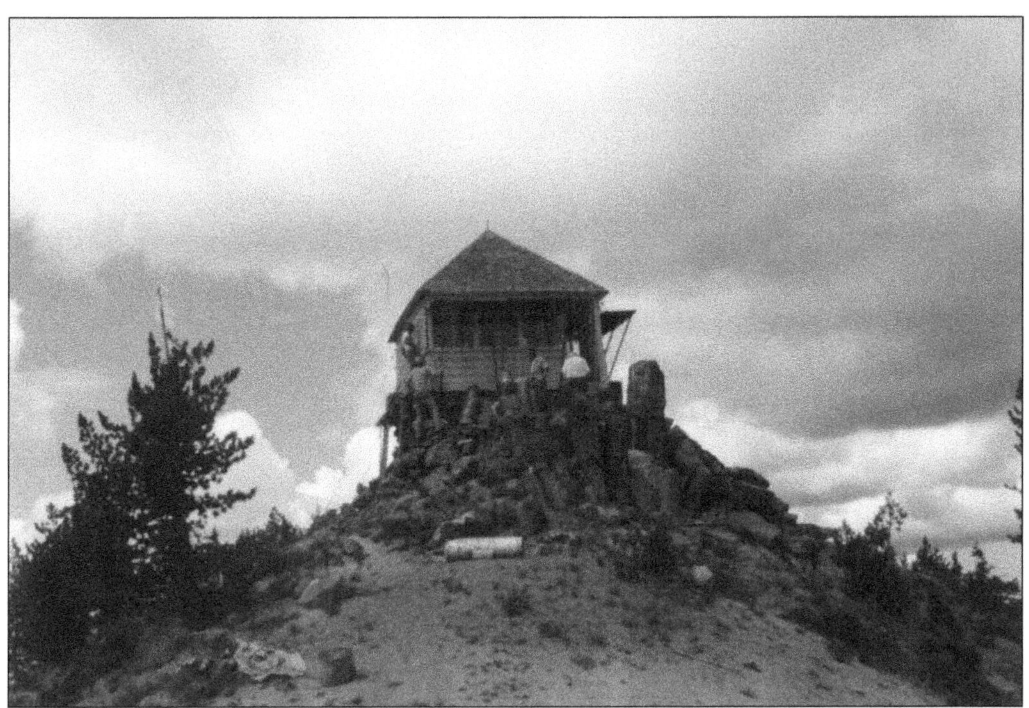

The D-5 lookout on 8,679-foot Mule Peak was built in 1924 and still stands, which makes it the second-oldest and second-highest standing lookout in Oregon. When Ronalee Leckenby was stationed here in 1984, she discovered upon arrival that one of the window panes was missing. Unable to get her supervisors to replace it in a timely manner, she covered it with four layers of plastic from a garbage bag. By September, she was glad for her patch job when a bitterly cold storm blew in, bringing snow and high winds.

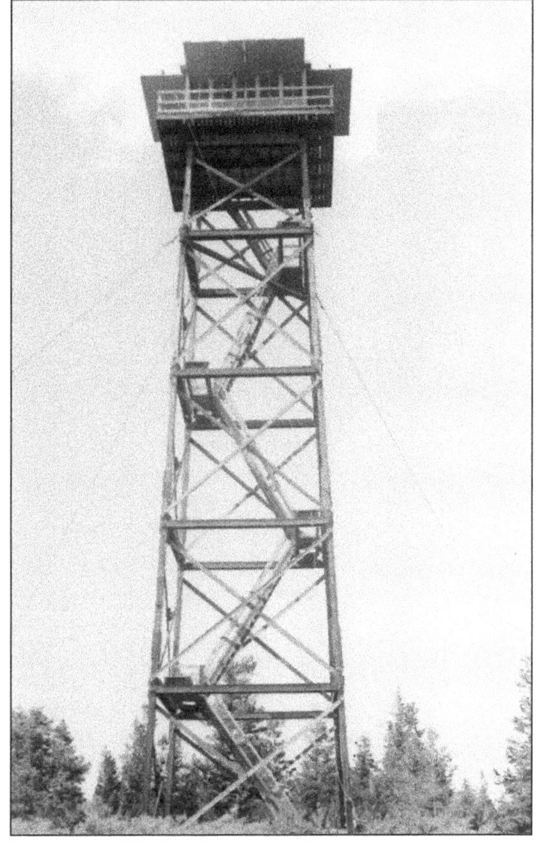

The first lookout on Arbuckle Mountain was a crow's nest that was built in the 1920s. It was followed by a 24-foot-tall platform tower in 1930. Pictured in 1962 is the 67-foot-tall L-4 tower that was built in 1946. It was torn down in 1976.

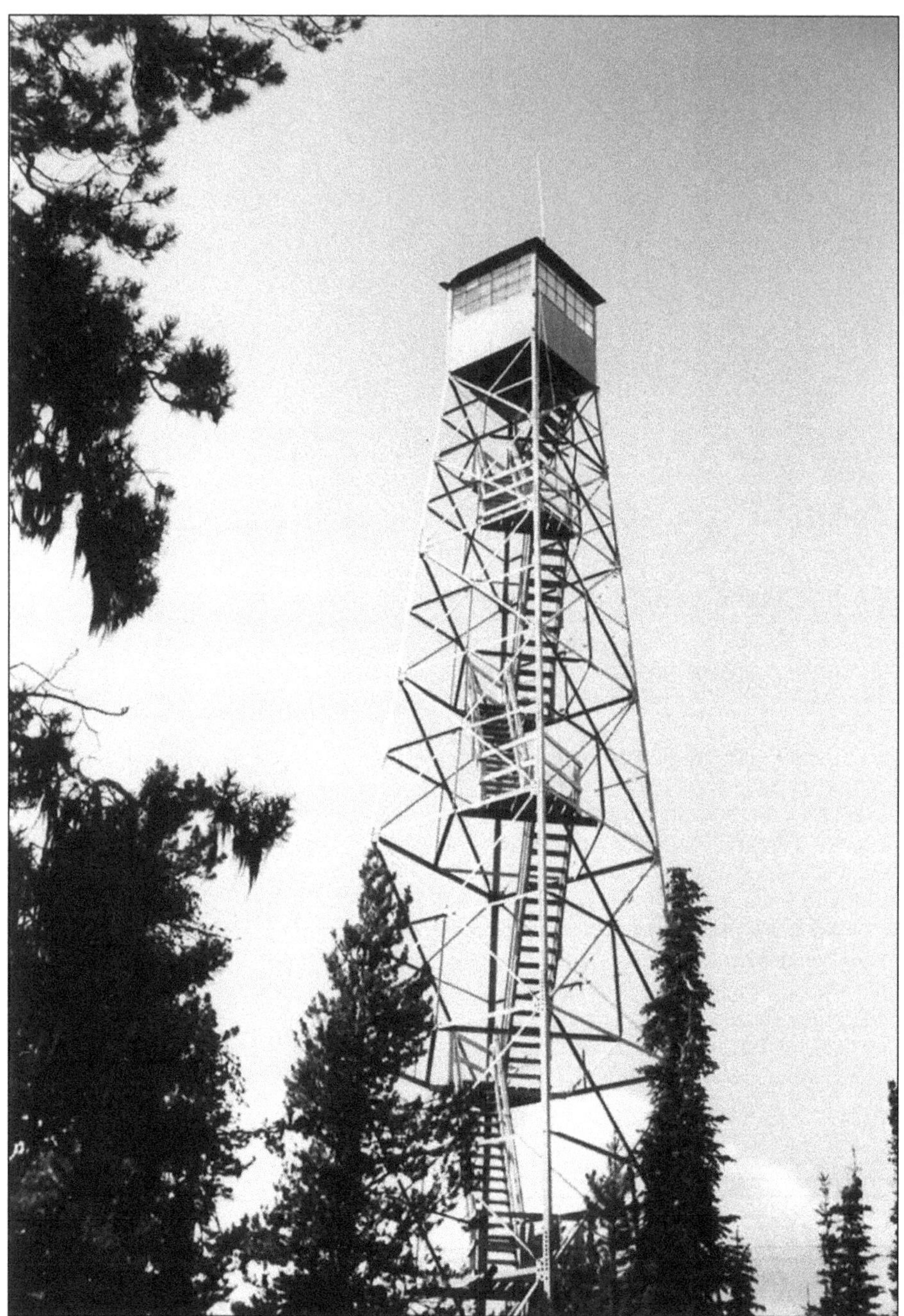

This 1957 photograph shows the 92-foot-tall steel tower that was built on Tower Mountain in 1935, replacing an earlier lookout that was built in the 1920s. The mountain had been known as Lookout Mountain, but its name was changed in 1925 because there was another Lookout Mountain nearby. In the late 1940s, the World War II veteran who was stationed there set fire to the ground cabin living quarters. Not realizing that arson was at play, the first people to respond to the fire were puzzled that they could not find him. While the unsuspecting residents from the nearby town of Ukiah led a search for him, he was in town looting their empty homes. The lookout is still staffed every summer.

This 40-foot-tall wooden lookout tower on Wheeler Point was built in 1936. This 1959 photograph shows the tower being lowered to the ground in preparation for its dismantling. A new tower replaced it but was removed in 1979.

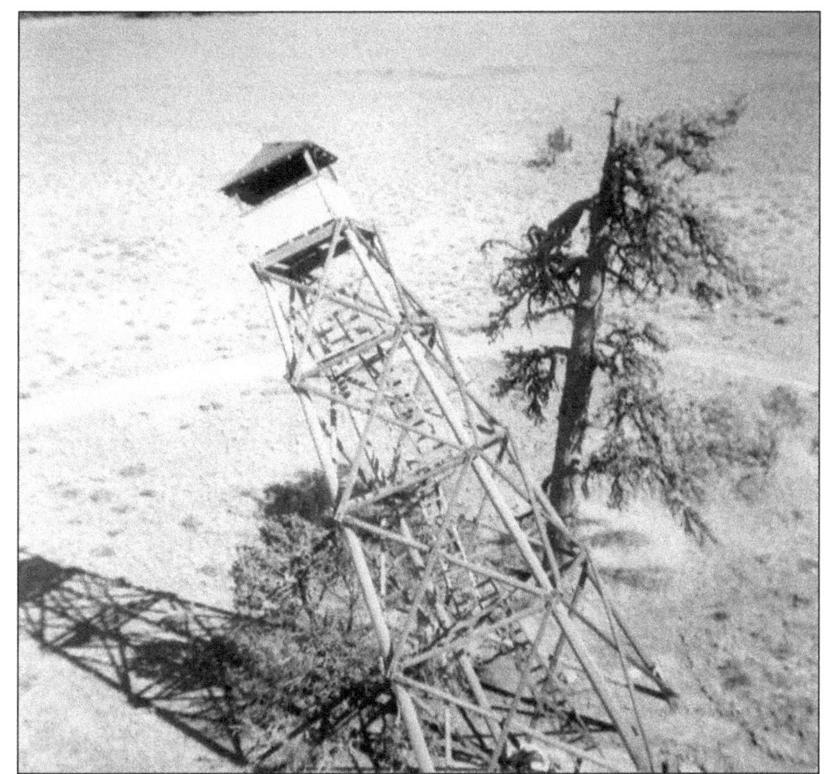

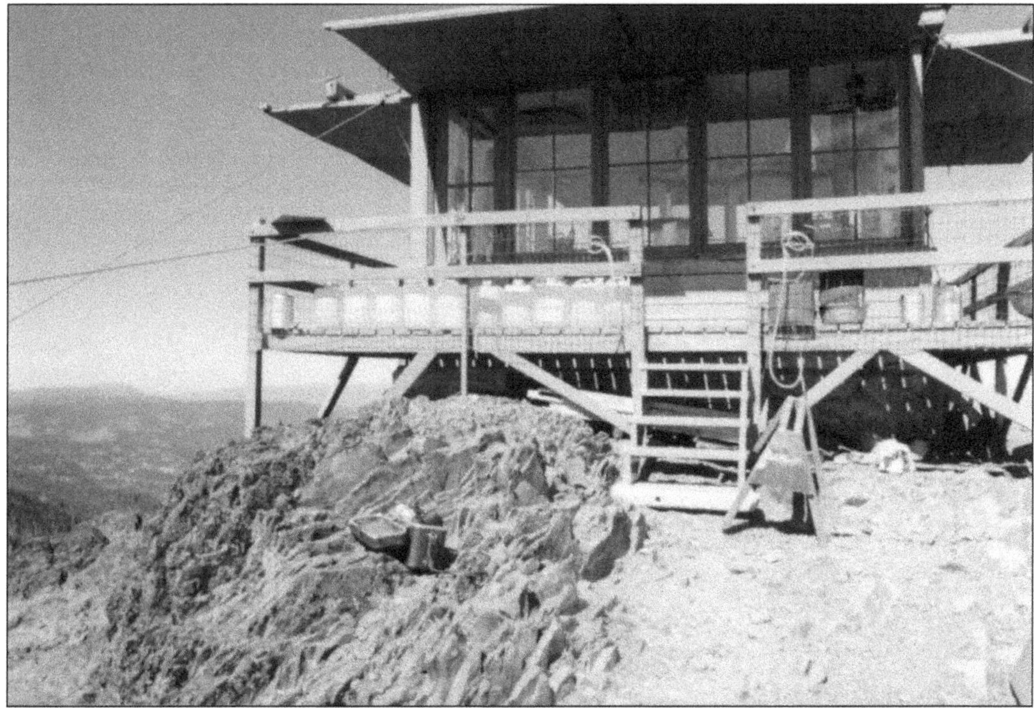

The first lookout on Table Rock was a cupola cabin built in the 1920s. Pictured is the second lookout, which was built in 1937. It is still staffed every summer.

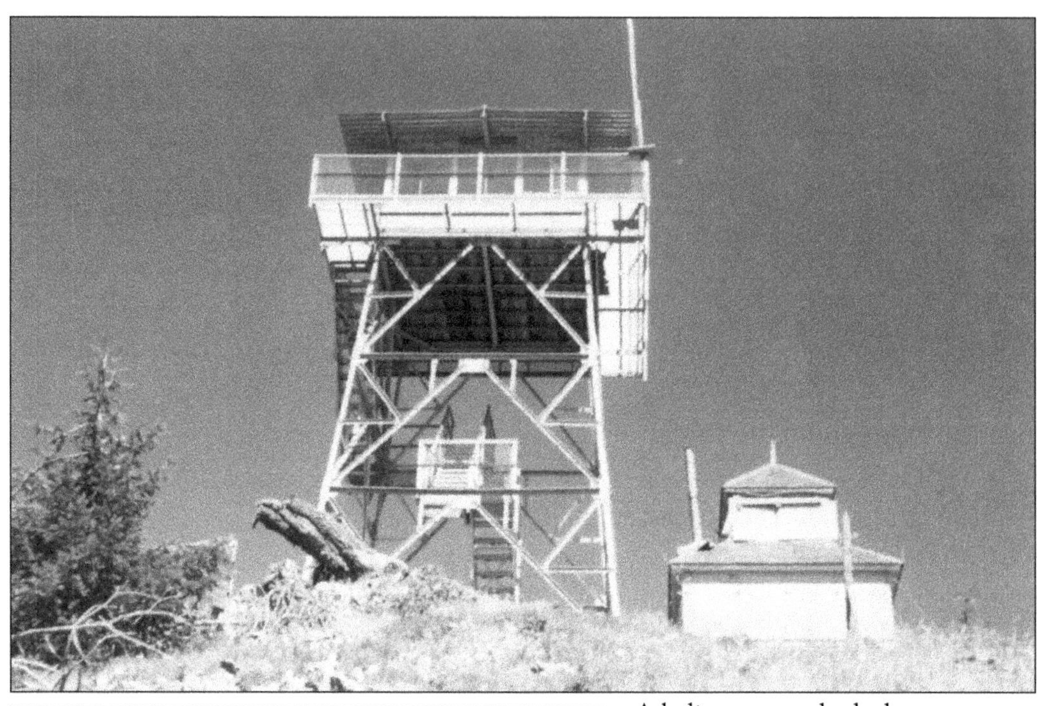

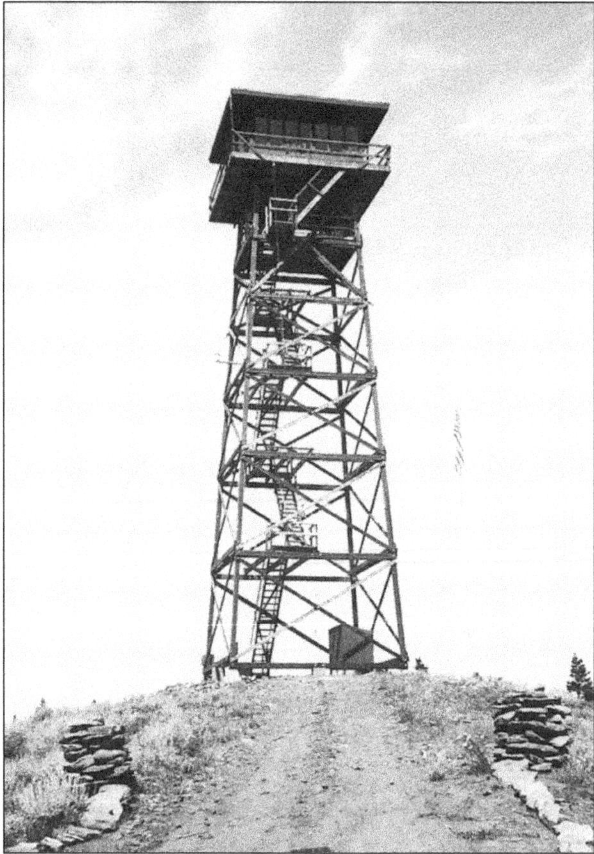

A helicopter crashed when attempting to land near the Madison Butte lookout in August 1981. While the helicopter was still 100 feet off the ground, the wind shifted and it fell to the ground. Although the helicopter was totaled, the four crew members escaped with minor injuries. The 37-foot-tall steel lookout tower, pictured in 1960, is still in use today.

This 1955 photograph shows the 24-foot-tall tower that was built on High Ridge in the Umatilla National Forest in 1941. It was replaced by a 67-foot-tall tower in 1959, which still stands today and is used in emergencies.

Royel and Lorene Allen were stationed at Ritter Butte in 1958. Lorene decided to learn to play the fiddle and ordered one from the Sears, Roebuck, and Company catalog. She started playing, but then "noticed something flying around the lookout, dodging under the shutters, trying to get in. It was an owl, out on the evening hunt, and from what he could hear, inside was a dying mouse. We looked at each other. I put the fiddle away." (Ron Kemnow.)

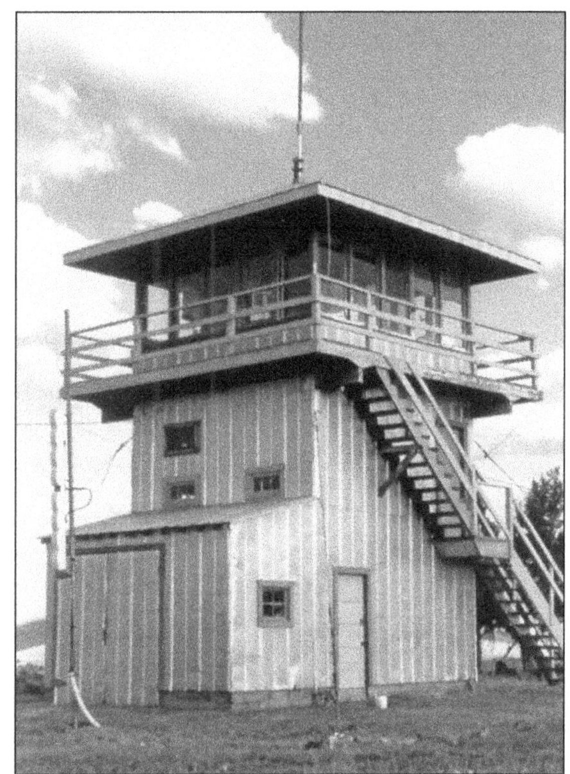

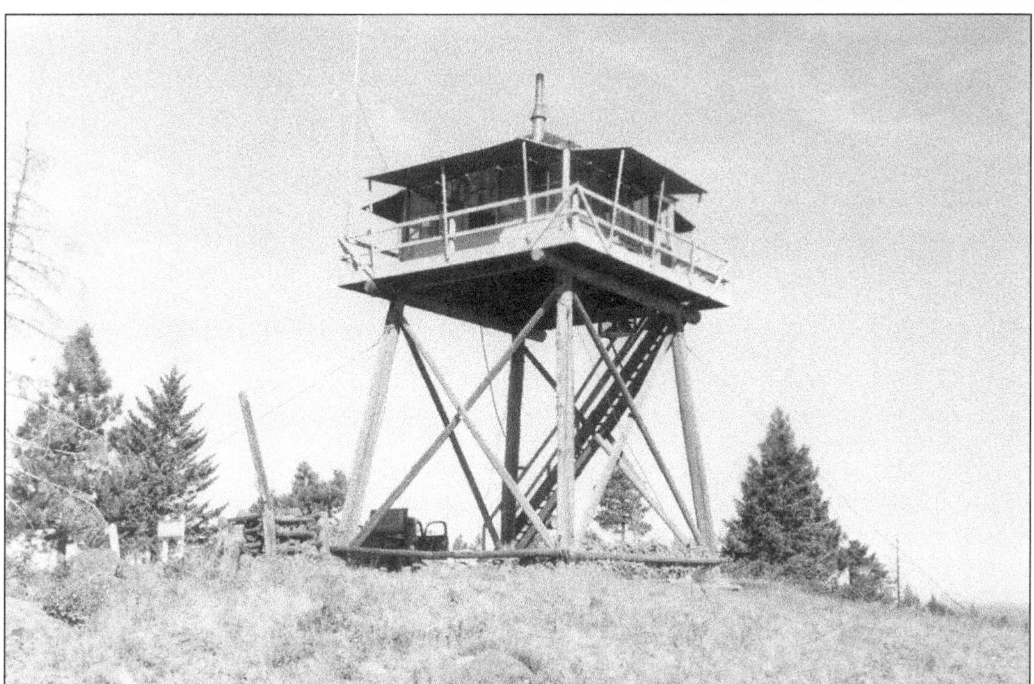

This 1961 photograph shows the 15-foot-tall Tiptop Mountain lookout in Umatilla County. It was built in the 1930s and replaced with a 40-foot-tall steel tower in 1960. The tower is no longer used and has been abandoned.

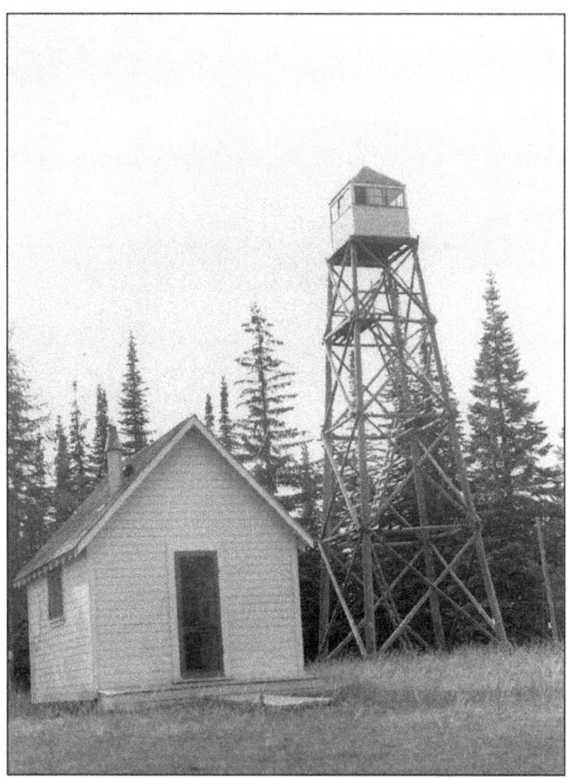

This lookout sits on one of many Lookout Mountains in Oregon. This mountain is in the Umatilla National Forest, and the 82-foot-tall tower was built in 1948. During refurbishment in 2004, the cab was lowered to the ground by crane and then raised back up again.

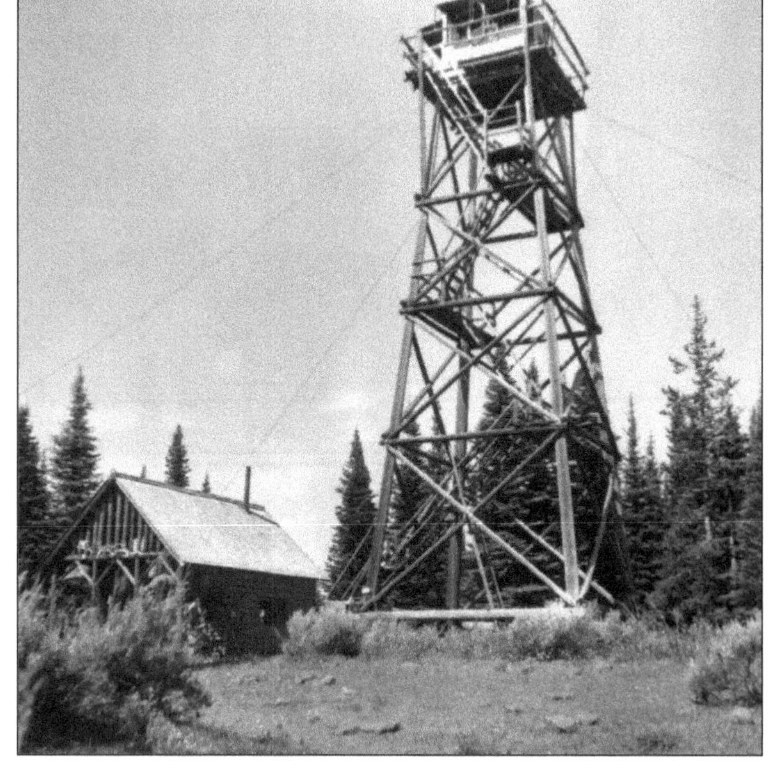

This Lookout Mountain is in the Wallowa-Whitman National Forest. The 50-foot-tall pole tower pictured was built in 1934 and later replaced by an 82-foot-tall tower in 1964. Just three years later, a forest fire destroyed the new tower, and it was never replaced.

This 1956 photograph shows the third and final Bone Spring lookout, which was located in the Umatilla National Forest on a high point above a spring with the same name. The spring was named for Chief Bones of the Palouse tribe. The 87-foot-tall lookout was removed in the 1980s.

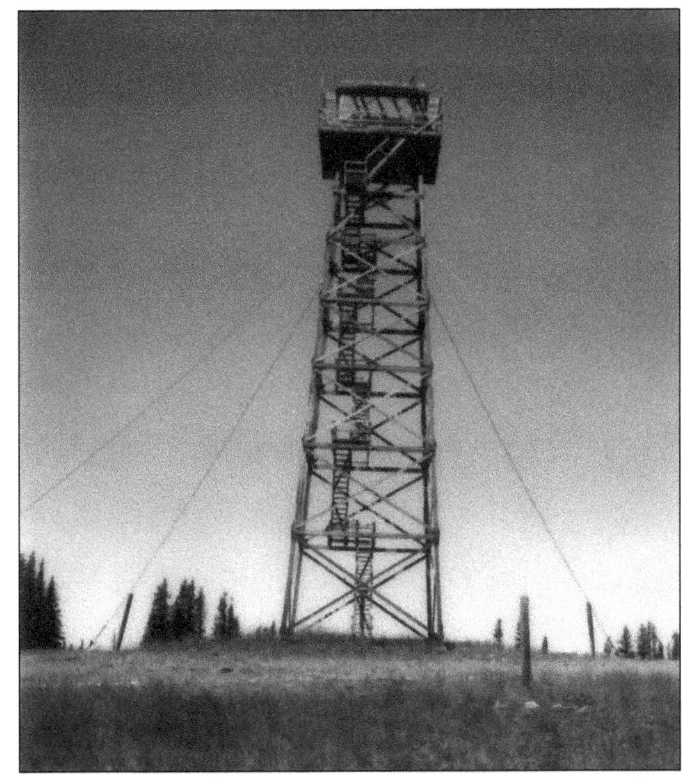

The tower pictured was built on Red Hill in 1936. In 1947, it was replaced with a new tower, which still stands today and is used for emergencies.

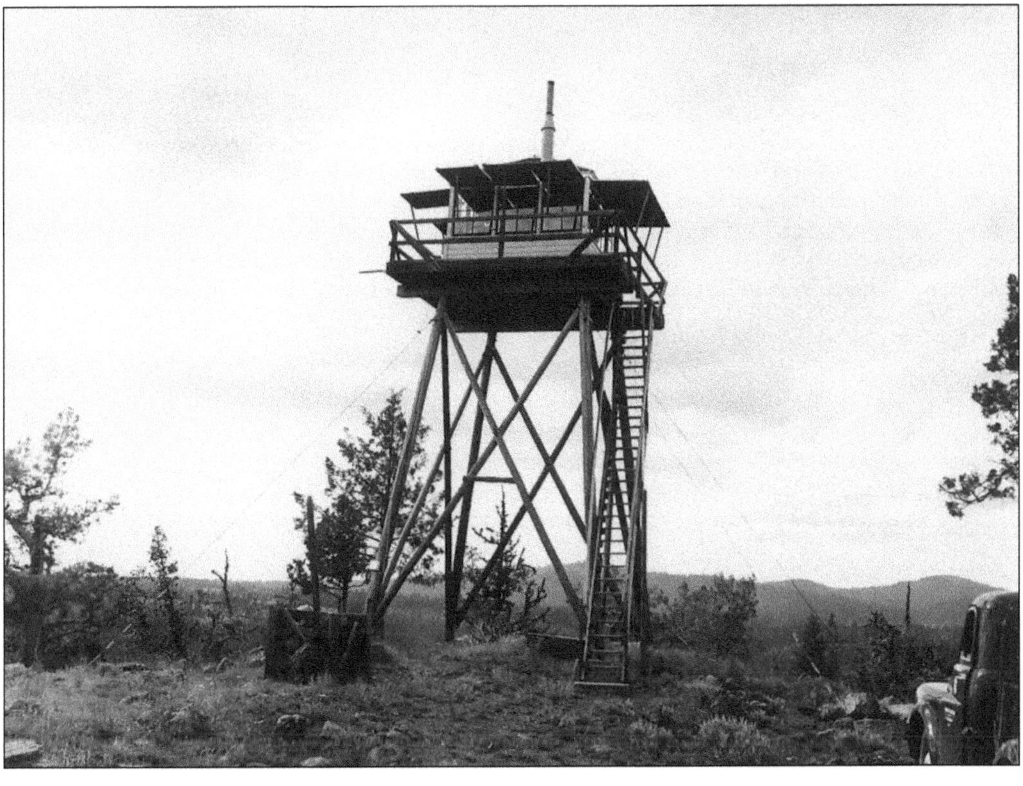

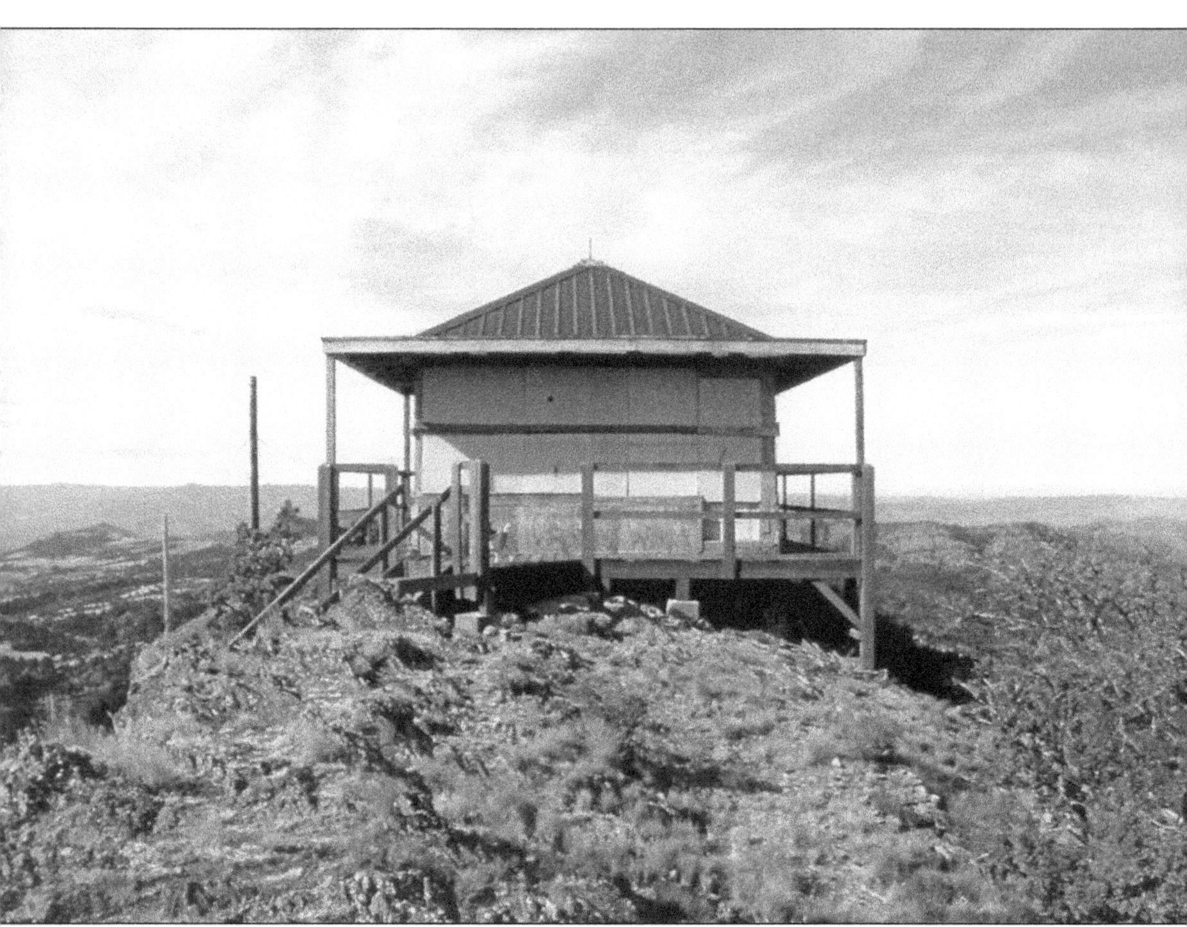

After two summers on Henkle Butte and a bout of rheumatic fever, Lois Christiansen Eagleton was stationed at Rancheria Rock in 1954 (also known as Rancherie Rock). For company she brought a caged parakeet, and also made friends with a ground squirrel she named "Goldie" who made himself at home in and around the cabin. He even tried, without success, to sample the sticky dough from a batch of cinnamon rolls left to rise on a window sill. Pictured in 2009, the lookout still stands, although it sits on private land and is no longer used. (Ron Kemnow.)

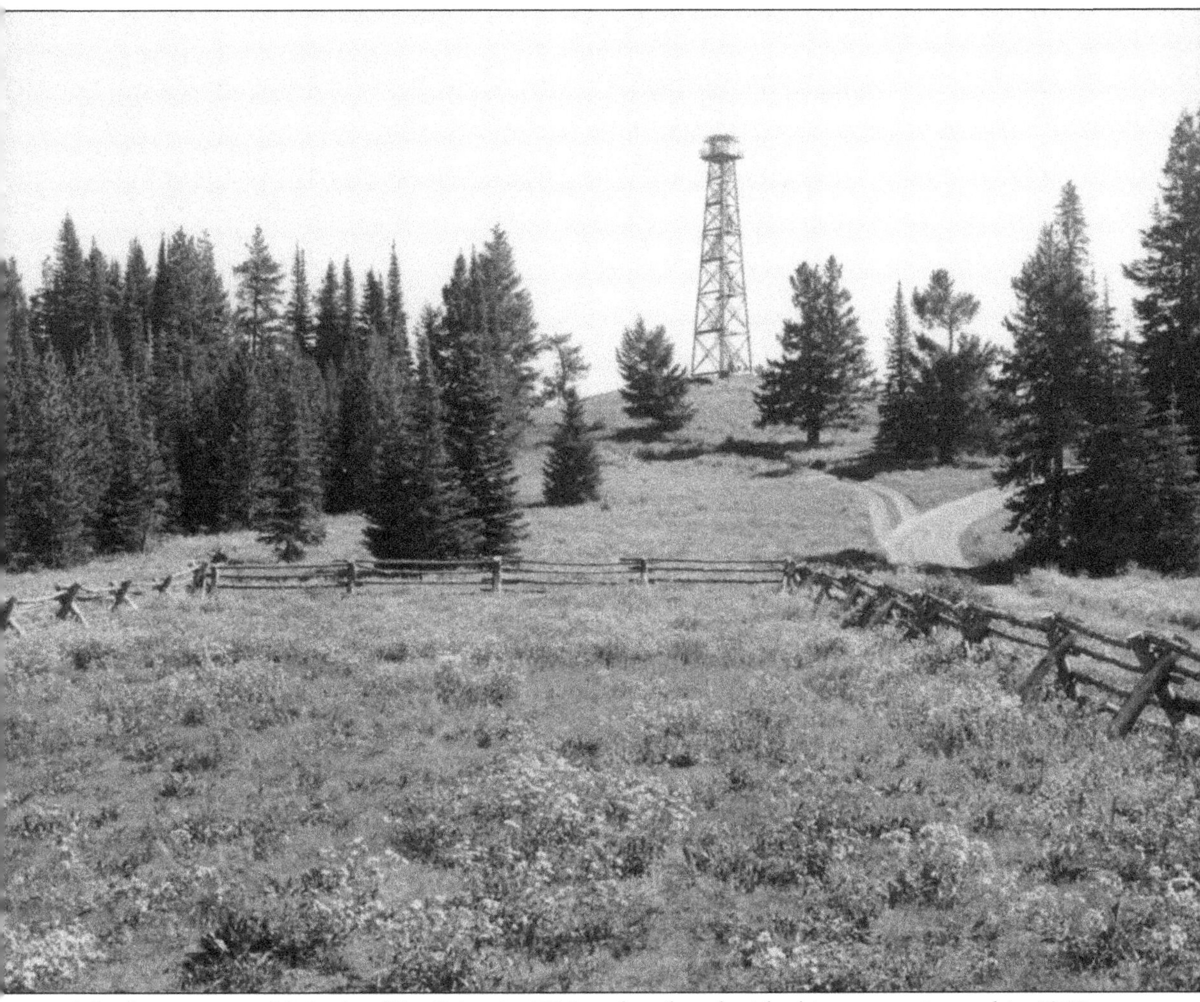

A lookout was established at Hat Point in 1916 and replaced with this tower, pictured in 1948. Later, an observation deck for visitor use was built halfway up the tower. It is still staffed every summer. In July 2007, C. Rod Bacon, who was staffing the lookout, had to evacuate when the Battle Creek Fire got too close for comfort. Aircraft were unable to drop fire retardant on the lookout, but a structure protection team was sent to do what they could from the ground. The fire got so close to the lookout that the roof caught flame and a three-foot-wide hole was burned in the observation deck; however, fire crews extinguished the flames, and the lookout was saved.

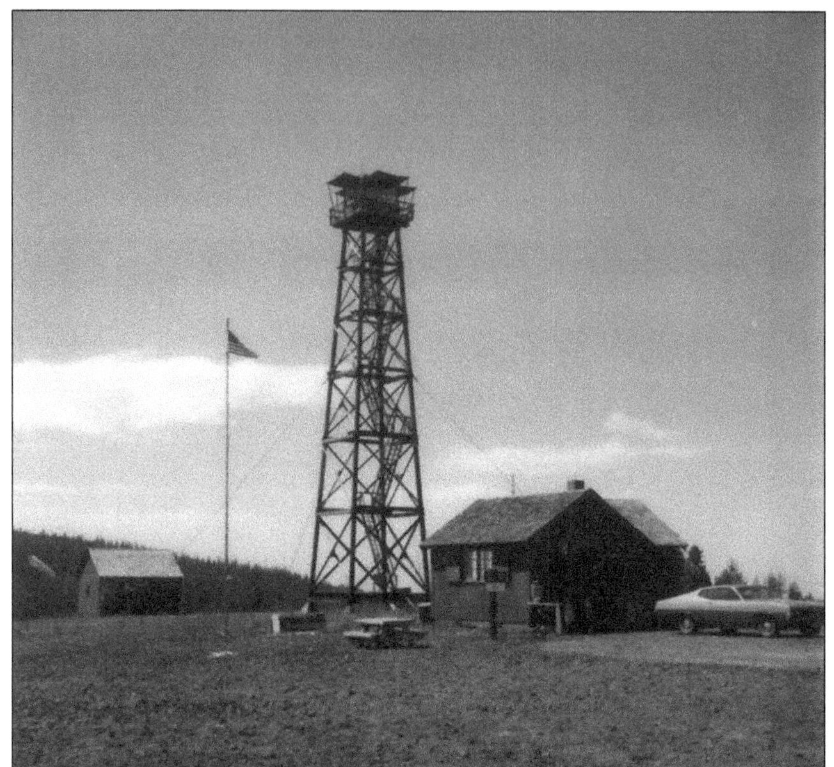

A 60-foot-tall tower was built on Goodman Ridge in 1936. It ceased to be staffed in the 1970s. The cabin blew off the top of the tower in a 2008 windstorm, and the ground cabin that had once served as living quarters was burned down the following year for removal.

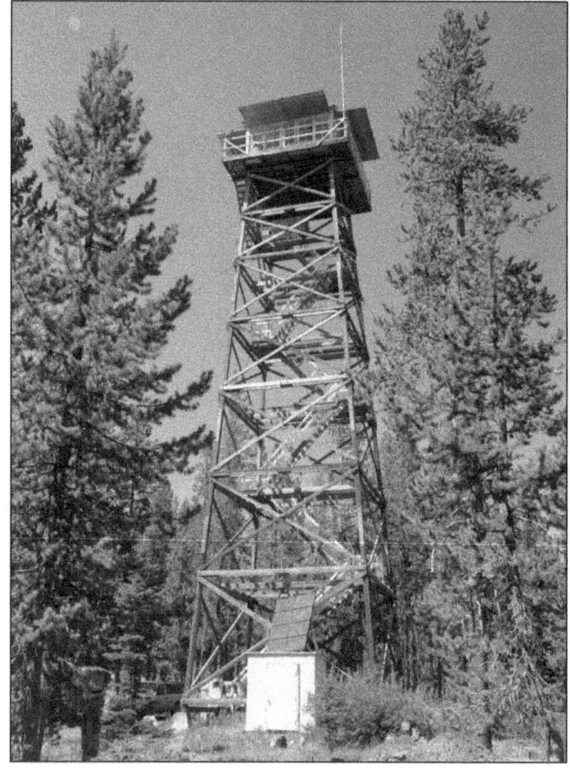

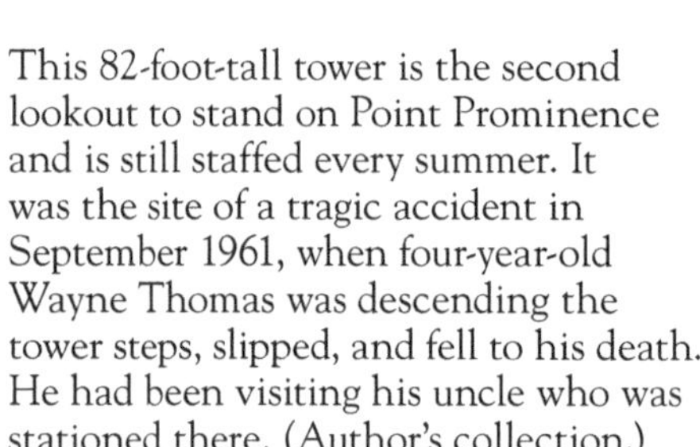

This 82-foot-tall tower is the second lookout to stand on Point Prominence and is still staffed every summer. It was the site of a tragic accident in September 1961, when four-year-old Wayne Thomas was descending the tower steps, slipped, and fell to his death. He had been visiting his uncle who was stationed there. (Author's collection.)

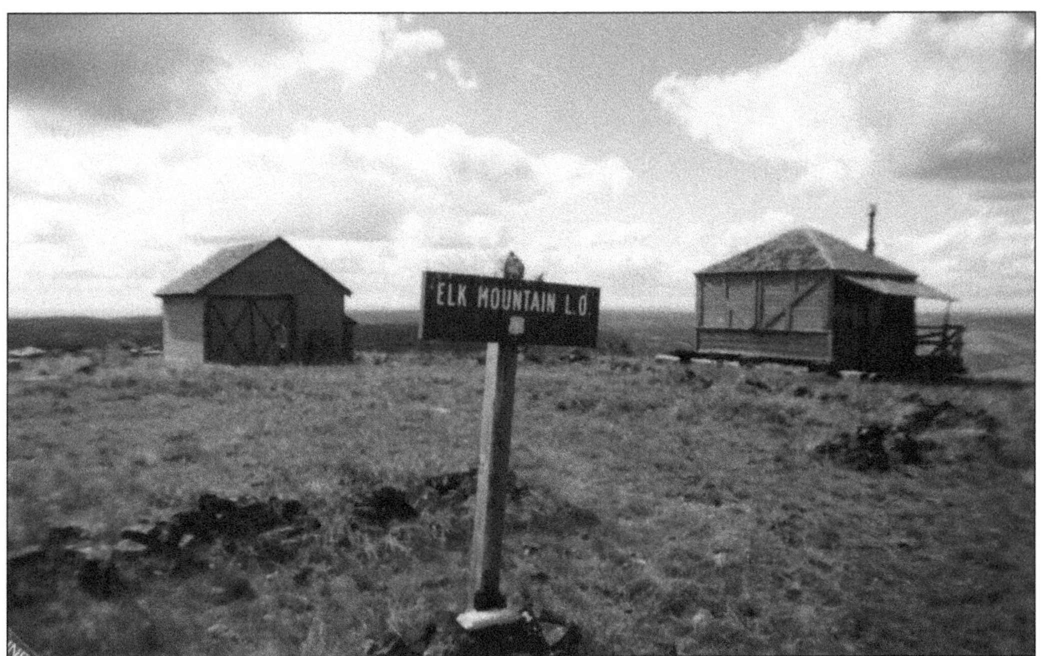

The lookout on Elk Mountain started out as a tower in 1933, but at some point, the tower was removed, and the cab was lowered to the ground. It was abandoned in the 1970s. The lookout and garage still stand, albeit on private property.

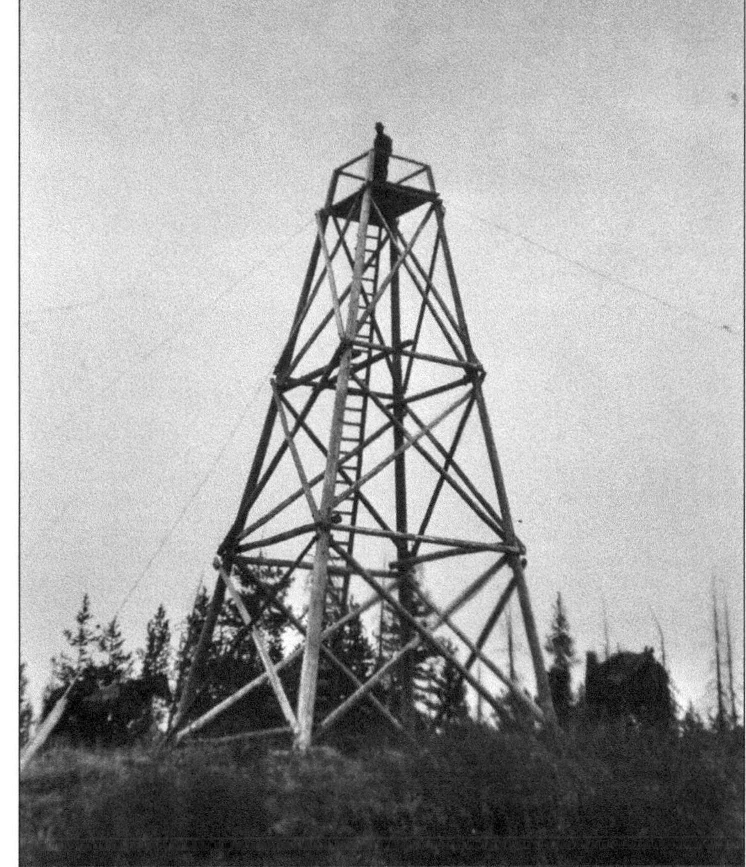

This was one of two platform towers that were erected on Beaver Ridge in the 1930s. The towers were half a mile apart and did not last long. They were abandoned in the 1940s.

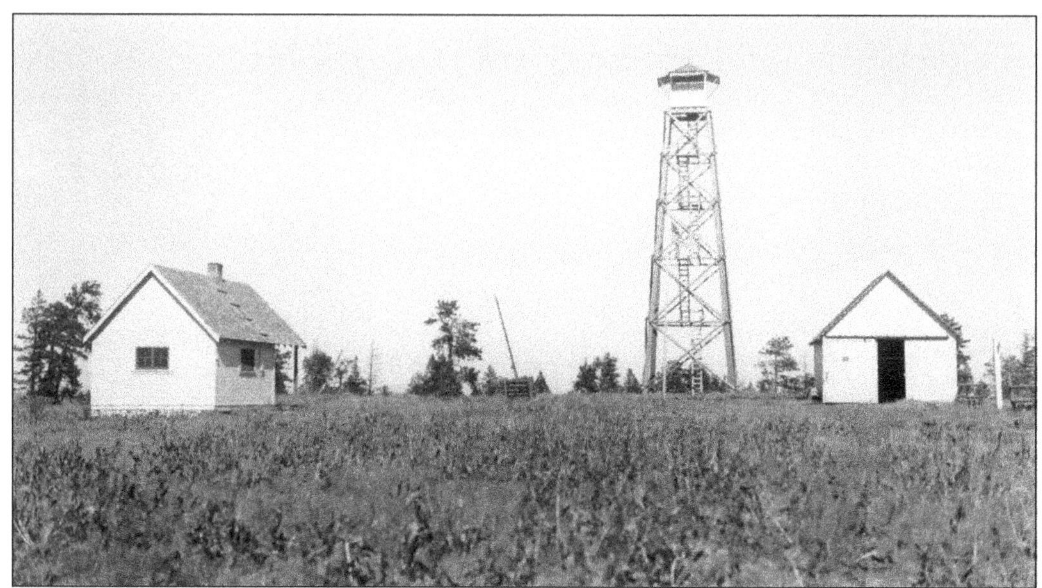

The 61-foot-tall tower pictured was built on Bally Mountain in 1933. It was replaced with an 82-foot-tall tower in 1960, which was removed in 1981.

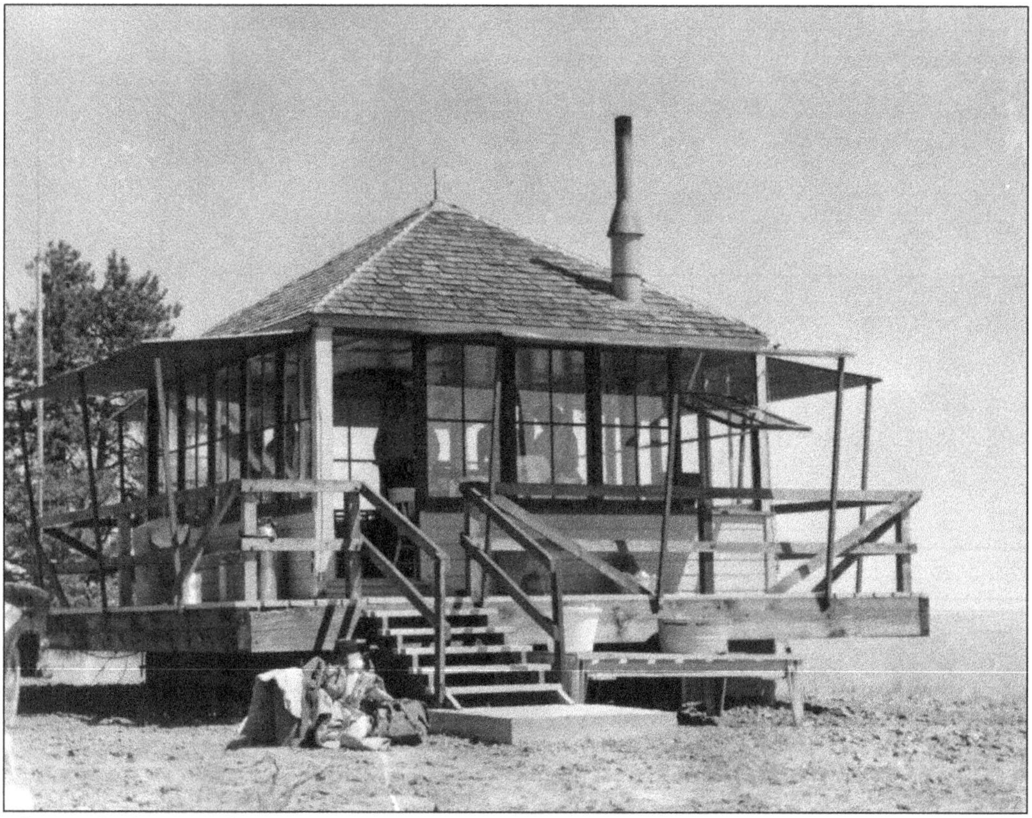

This L-4 lookout was built on Buckhorn Mountain in 1947. It has not been staffed since the 1970s, and the lookout is abandoned. The site is one of three viewpoints overlooking Hells Canyon, so it still receives visitors.

The first lookout on Mount Harris was built in 1930, and the tower pictured was built in 1957. It is no longer staffed but remains standing as a base for radio equipment. (Author's collection.)

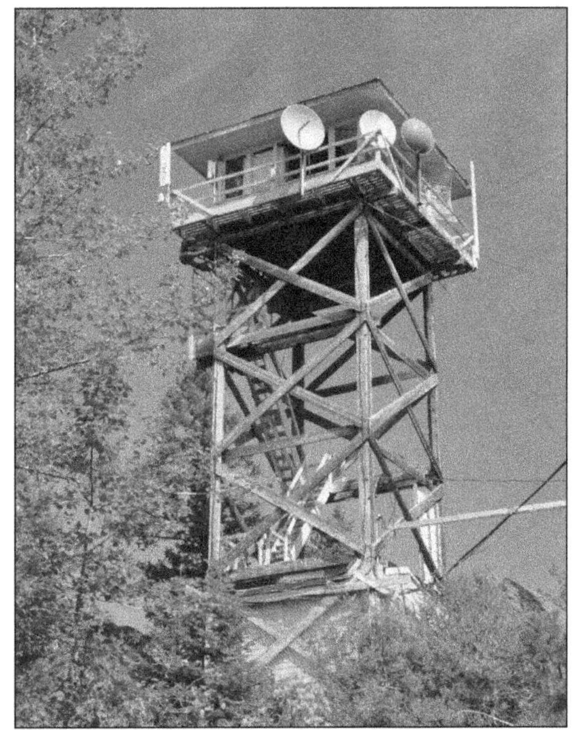

A 50-foot-tall tower was built on Russell Mountain in 1921. It was replaced by an 82-foot-tall tower with an open platform in 1941. The platform was enclosed by an L-4 cab in 1950. The tower came close to burning down in 1994 when the Twin Lakes Fire came within 100 yards of the lookout.

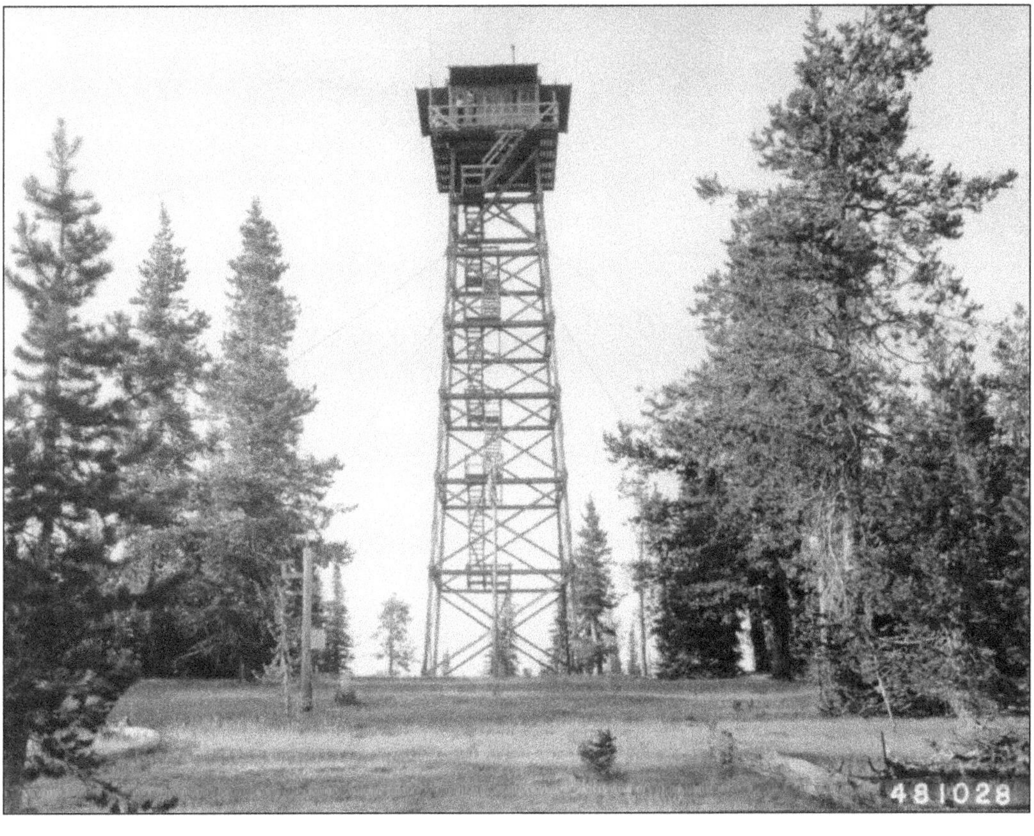

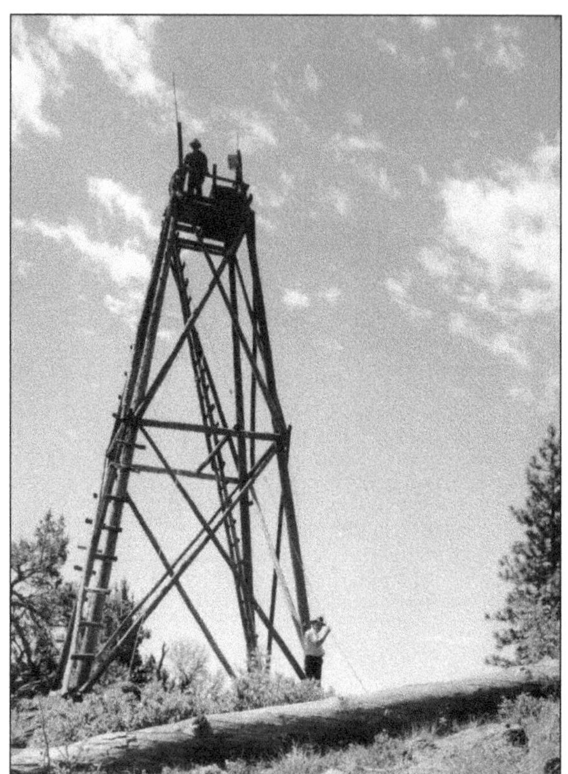

The 43-foot-tall platform tower was built on Pogue Point in the 1920s. In 1934, it was replaced by a 22-foot-tall L-4 tower, which was then replaced by a 41-foot-tall tower. It ceased to be used by the 1960s and was removed.

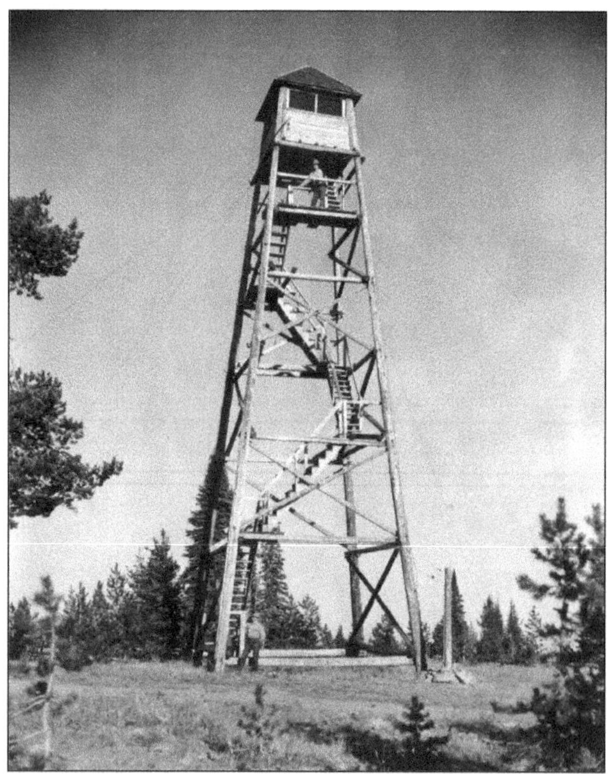

A 68-foot-tall lookout was built at Trinity Guard Station in 1926. It was located on an unnamed high point one mile south of Fish Lake. It was removed in 1953.

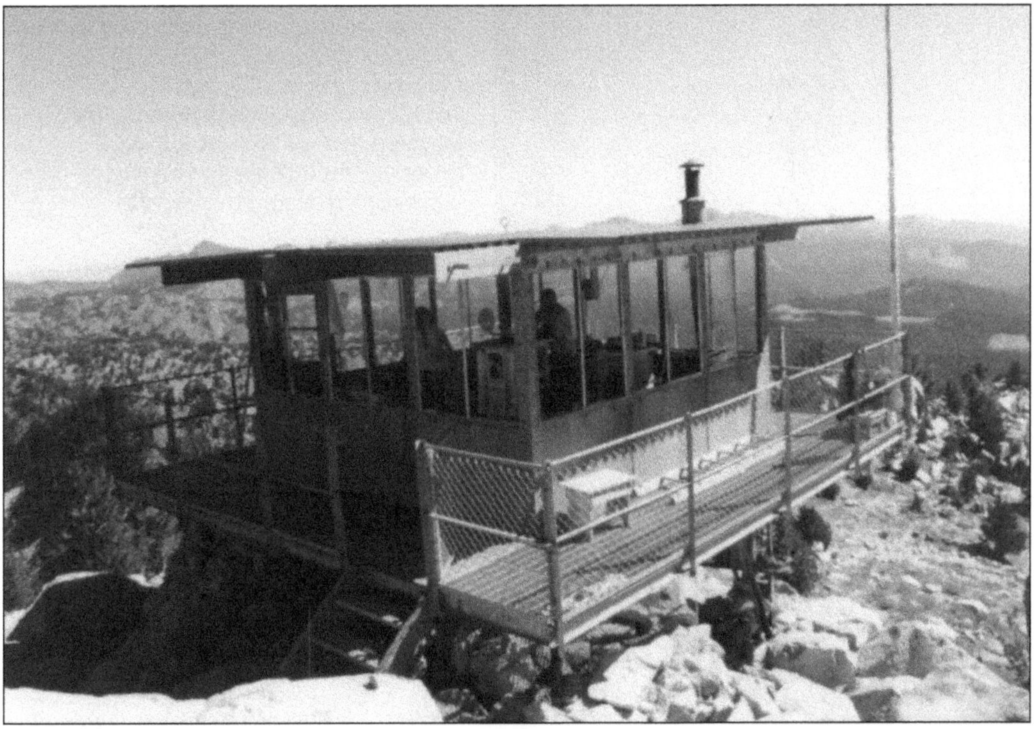

Weather on a mountaintop can be harsh, even in summer. When Deanna Fox and Dan Wilcock were staffing the lookout on 8,346-foot Mount Ireland in 2000, they experienced a severe summer ice storm. They had to pry eight-inch-long ice crystals from the radio antenna to keep it from toppling. In September of that same year, the temperature on the summit was 10 degrees Fahrenheit one night.

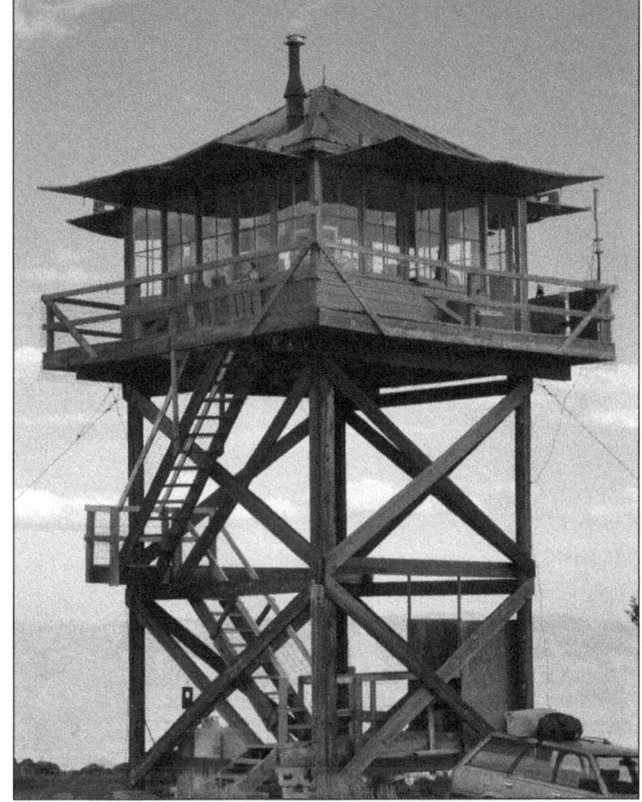

The lookout on Summit Point was built in 1951 and is still staffed every summer. When lookout Carolanna Savage was interviewed by a reporter in 2012, she said she was destined for such a job. Her mother, Joanna Harris Rode, was pregnant with her when working a lookout tower. Savage brings her dog Queen to the lookout. "Don't ever name a dog Queen, because it goes to her head," she told the reporter. (Author's collection.)

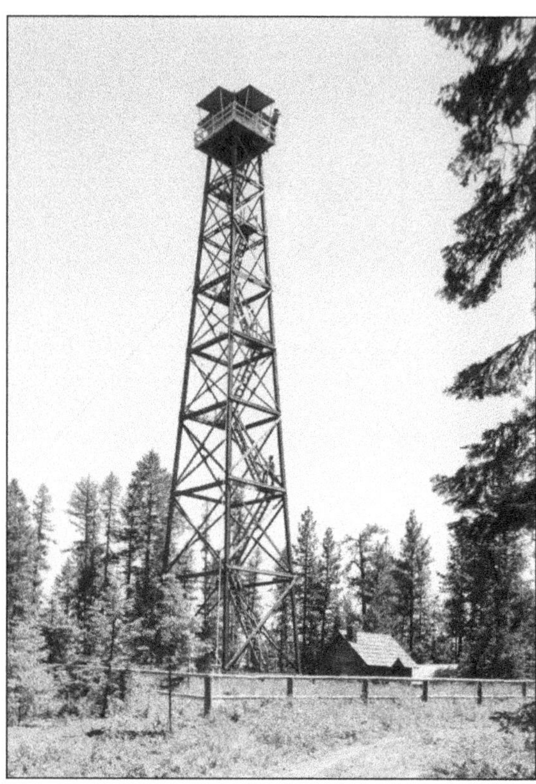

A crow's nest tree lookout was established on Kirkland Butte in 1926. A 100-foot-tall tower (pictured) replaced it in 1934, but a strong windstorm in 1940 blew it down. A 101-foot-tall steel tower was built the following year but was torn down in the 1950s.

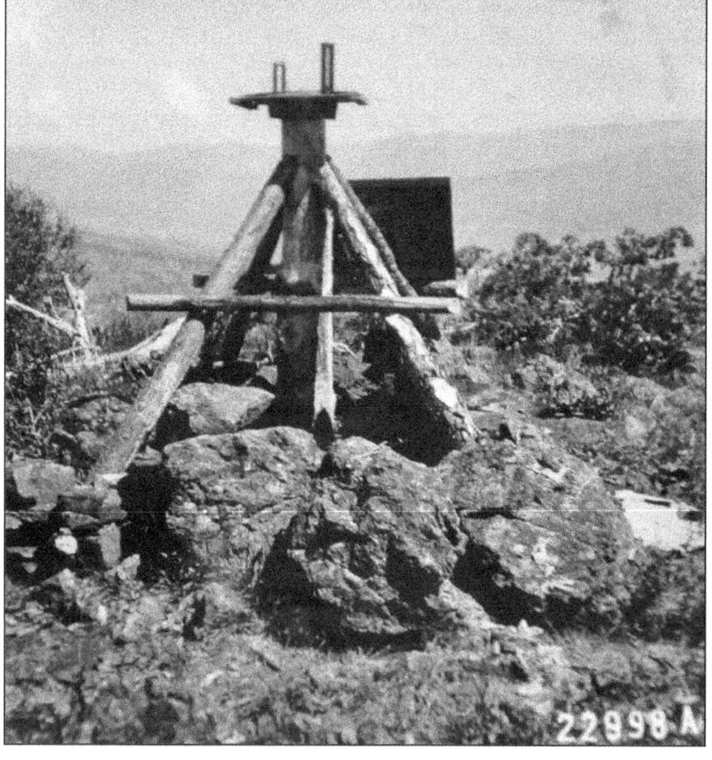

Although the essential job of lookouts has not changed, the living conditions have. While lookout sites started out as simple stump-mounted fire finders, like this one on Huckleberry Mountain, today's lookouts have propane heat, solar power, and radio communication.

Bibliography

Anderson, Rolf, ed. *We Had an Objective in Mind: The US Forest Service in the Pacific Northwest 1905 to 2005*. Portland, OR: Pacific Northwest Forest Service Association, 2005.

Arnst, Albert. *We Climbed the Highest Mountains*. Portland, OR: A. Arnst, 1985.

Davies, Gilbert W. and Florice M. Frank, eds. *Forest Service Memories: Past Lives and Times in the United States Forest Service*. Hat Creek, CA: History Ink Books, 1997.

Kresek, Ray. *Fire Lookouts of the Northwest*. Fairfield, WA: Ye Galleon Press, 1984.

Lewis, James G. *The Forest Service and the Greatest Good: A Centennial History*. Durham, NC: Forest History Society, 2005.

Lucia, Ellis. *Tillamook Burn Country: A Pictorial History*. Caldwell, ID: Caxton Printers, 1983.

McArthur, Lewis A. *Oregon Geographic Names*. Portland, OR: Oregon Historical Society Press, 2003.

Newman, Doug. *Finding Fire: A Personal History of Fire Lookouts in Lane County, Oregon*. Eugene, OR: Lane County Historical Society, 1982.

Suiter, John. *Poets on the Peaks: Jack Snyder, Philip Whalen & Jack Kerouac in the North Cascades*. Washington, DC: Counterpoint, 2002.

Williams, Gerald W. *The US Forest Service in the Pacific Northwest: A History*. Corvallis, OR: Oregon State University Press, 2009.

www.ingramcontent.com/pod-product-compliance
Lightning Source LLC
LaVergne TN
LVHW060625110826
845147LV00015B/938

* 9 7 8 1 4 6 7 1 3 4 8 6 6 *